HOLT SCIENCE & TECHNOLOGY

Earth's Changing Surface

HOLT, RINEHART AND WINSTON

A Harcourt Classroom Education Company

Austin · New York · Orlando · Atlanta · San Francisco · Boston · Dallas · Toronto · London

Staff Credits

Editorial

Robert W. Todd, Executive Editor

Robert V. Tucek, Leigh Ann Garcia, Senior Editors

Clay Walton, Jim Ratcliffe, Editors

ANCILLARIES

Jennifer Childers, Senior Editor

Chris Colby, Molly Frohlich, Shari Husain, Kristen McCardel, Sabelyn Pussman, Erin Roberson

COPYEDITING

Dawn Spinozza, Copyediting Supervisor

EDITORIAL SUPPORT STAFF

Jeanne Graham, Mary Helbling, Tanu'e White, Doug Rutley

EDITORIAL PERMISSIONS

Cathy Paré, Permissions Manager

Jan Harrington, Permissions Editor

Art, Design, and Photo

BOOK DESIGN

Richard Metzger, Design Director

Marc Cooper, Senior Designer

José Garza, Designer

Alicia Sullivan, Designer (ATE), **Cristina Bowerman**, Design Associate (ATE), **Eric Rupprath**, Designer (Ancillaries), **Holly Whittaker**, Traffic Coordinator

IMAGE ACQUISITIONS

Joe London, Director

Elaine Tate, Art Buyer Supervisor

Jeannie Taylor, Photo Research Supervisor

Andy Christiansen, Photo Researcher

Jackie Berger, Assistant Photo Researcher

PHOTO STUDIO

Sam Dudgeon, Senior Staff Photographer

Victoria Smith, Photo Specialist

Lauren Eischen, Photo Coordinator

DESIGN NEW MEDIA

Susan Michael, Design Director

Production

Mimi Stockdell, Senior Production Manager

Beth Sample, Senior Production Coordinator

Suzanne Brooks, Sara Carroll-Downs

Media Production

Kim A. Scott, Senior Production Manager

Adriana Bardin-Prestwood, Senior Production Coordinator

New Media

Armin Gutzmer, Director

Jim Bruno, Senior Project Manager

Lydia Doty, Senior Project Manager

Jessica Bega, Project Manager

Cathy Kuhles, Nina Degollado, Technical Assistants

Design Implementation and Production

The Quarasan Group, Inc.

Acknowledgments

Chapter Writers

Kathleen Meehan Berry
Science Chairman
Canon-McMillan School District
Canonsburg, Pennsylvania

Robert H. Fronk, Ph.D.
Chair of Science and Mathematics Education Department
Florida Institute of Technology
West Melbourne, Florida

Mary Kay Hemenway, Ph.D.
Research Associate and Senior Lecturer
Department of Astronomy
The University of Texas
Austin, Texas

Kathleen Kaska
Life and Earth Science Teacher
Lake Travis Middle School
Austin, Texas

Peter E. Malin, Ph.D.
Professor of Geology
Division of Earth and Ocean Sciences
Duke University
Durham, North Carolina

Karen J. Meech, Ph.D.
Associate Astronomer
Institute for Astronomy
University of Hawaii
Honolulu, Hawaii

Robert J. Sager
Chair and Professor of Earth Sciences
Pierce College
Lakewood, Washington

Lab Writers

Kenneth Creese
Science Teacher
White Mountain Junior High School
Rock Springs, Wyoming

Linda A. Culp
Science Teacher and Dept. Chair
Thorndale High School
Thorndale, Texas

Bruce M. Jones
Science Teacher and Dept. Chair
The Blake School
Minneapolis, Minnesota

Shannon Miller
Science and Math Teacher
Llano Junior High School
Llano, Texas

Robert Stephen Ricks
Special Services Teacher
Department of Classroom Improvement
Alabama State Department of Education
Montgomery, Alabama

James J. Secosky
Science Teacher
Bloomfield Central School
Bloomfield, New York

Academic Reviewers

Mead Allison, Ph.D.
Assistant Professor of Oceanography
Texas A&M University
Galveston, Texas

Alissa Arp, Ph.D.
Director and Professor of Environmental Studies
Romberg Tiburon Center
San Francisco State University
Tiburon, California

Paul D. Asimow, Ph.D.
Assistant Professor of Geology and Geochemistry
Department of Physics and Planetary Sciences
California Institute of Technology
Pasadena, California

G. Fritz Benedict, Ph.D.
Senior Research Scientist and Astronomer
McDonald Observatory
The University of Texas
Austin, Texas

Russell M. Brengelman, Ph.D.
Professor of Physics
Morehead State University
Morehead, Kentucky

John A. Brockhaus, Ph.D.
Director—Mapping, Charting, and Geodesy Program
Department of Geography and Environmental Engineering
United States Military Academy
West Point, New York

Michael Brown, Ph.D.
Assistant Professor of Planetary Astronomy
Department of Physics and Astronomy
California Institute of Technology
Pasadena, California

Wesley N. Colley, Ph.D.
Postdoctoral Fellow
Harvard-Smithsonian Center for Astrophysics
Cambridge, Massachusetts

Andrew J. Davis, Ph.D.
Manager—ACE Science Data Center
Physics Department
California Institute of Technology
Pasadena, California

Peter E. Demmin, Ed.D.
Former Science Teacher and Department Chair
Amherst Central High School
Amherst, New York

James Denbow, Ph.D.
Associate Professor
Department of Anthropology
The University of Texas
Austin, Texas

Roy W. Hann, Jr., Ph.D.
Professor of Civil Engineering
Texas A&M University
College Station, Texas

Frederick R. Heck, Ph.D.
Professor of Geology
Ferris State University
Big Rapids, Michigan

Richard Hey, Ph.D.
Professor of Geophysics
Hawaii Institute of Geophysics and Planetology
University of Hawaii
Honolulu, Hawaii

John E. Hoover, Ph.D.
Associate Professor of Biology
Millersville University
Millersville, Pennsylvania

Robert W. Houghton, Ph.D.
Senior Staff Associate
Lamont-Doherty Earth Observatory
Columbia University
Palisades, New York

Steven A. Jennings, Ph.D.
Assistant Professor
Department of Geography & Environmental Studies
University of Colorado
Colorado Springs, Colorado

Eric L. Johnson, Ph.D.
Assistant Professor of Geology
Central Michigan University
Mount Pleasant, Michigan

John Kermond, Ph.D.
Visiting Scientist
NOAA–Office of Global Programs
Silver Spring, Maryland

Zavareh Kothavala, Ph.D.
Postdoctoral Associate Scientist
Department of Geology and Geophysics
Yale University
New Haven, Connecticut

Karen Kwitter, Ph.D.
Ebenezer Fitch Professor of Astronomy
Williams College
Williamstown, Massachusetts

Valerie Lang, Ph.D.
Project Leader of Environmental Programs
The Aerospace Corporation
Los Angeles, California

Philip LaRoe
Professor
Helena College of Technology
Helena, Montana

Julie Lutz, Ph.D.
Astronomy Program
Washington State University
Pullman, Washington

Duane F. Marble, Ph.D.
Professor Emeritus
Department of Geography and Natural Resources
Ohio State University
Columbus, Ohio

Joseph A. McClure, Ph.D.
Associate Professor
Department of Physics
Georgetown University
Washington, D.C.

Frank K. McKinney, Ph.D.
Professor of Geology
Appalachian State University
Boone, North Carolina

Joann Mossa, Ph.D.
Associate Professor of Geography
University of Florida
Gainesville, Florida

LaMoine L. Motz, Ph.D.
Coordinator of Science Education
Department of Learning Services
Oakland County Schools
Waterford, Michigan

Barbara Murck, Ph.D.
Assistant Professor of Earth Science
Erindale College
University of Toronto
Mississauga, Ontario, Canada

Hilary Clement Olson, Ph.D.
Research Associate
Institute for Geophysics
The University of Texas
Austin, Texas

Andre Potochnik
Geologist
Grand Canyon Field Institute
Flagstaff, Arizona

John R. Reid, Ph.D.
Professor Emeritus
Department of Geology and Geological Engineering
University of North Dakota
Grand Forks, North Dakota

Gary Rottman, Ph.D.
Associate Director
Laboratory for Atmosphere and Space Physics
University of Colorado
Boulder, Colorado

Dork L. Sahagian, Ph.D.
Professor
Institute for the Study of Earth, Oceans, and Space
University of New Hampshire
Durham, New Hampshire

Peter Sheridan, Ph.D.
Professor of Chemistry
Colgate University
Hamilton, New York

David Sprayberry, Ph.D.
Assistant Director for Observing Support
W.M. Keck Observatory
California Association for Research in Astronomy
Kamuela, Hawaii

Lynne Talley, Ph.D.
Professor
Scripps Institution of Oceanography
University of California
La Jolla, California

Acknowledgments (cont.)

Glenn Thompson, Ph.D.
Scientist
Geophysical Institute
University of Alaska
Fairbanks, Alaska

Martin VanDyke, Ph.D.
Professor of Chemistry, Emeritus
Front Range Community
College
Westminister, Colorado

Thad A. Wasklewicz, Ph.D.
Assistant Professor of Geography
University of Memphis
Memphis, Tennessee

Hans Rudolf Wenk, Ph.D.
*Professor of Geology and
Geophysical Sciences*
University of California
Berkeley, California

Lisa D. White, Ph.D.
Associate Professor of Geosciences
San Francisco State University
San Francisco, California

Lorraine W. Wolf, Ph.D.
Associate Professor of Geology
Auburn University
Auburn, Alabama

Charles A. Wood, Ph.D.
*Chairman and Professor of Space
Studies*
University of North Dakota
Grand Forks, North Dakota

Safety Reviewer

Jack Gerlovich, Ph.D.
Associate Professor
School of Education
Drake University
Des Moines, Iowa

Teacher Reviewers

Barry L. Bishop
Science Teacher and Dept. Chair
San Rafael Junior High School
Ferron, Utah

Yvonne Brannum
Science Teacher and Dept. Chair
Hine Junior High School
Washington, D.C.

Daniel L. Bugenhagen
Science Teacher and Dept. Chair
Yutan Junior & Senior High
School
Yutan, Nebraska

Kenneth Creese
Science Teacher
White Mountain Junior High
School
Rock Springs, Wyoming

Linda A. Culp
Science Teacher and Dept. Chair
Thorndale High School
Thorndale, Texas

Alonda Droege
Science Teacher
Pioneer Middle School
Steilacom, Washington

Laura Fleet
Science Teacher
Alice B. Landrum Middle
School
Ponte Vedra Beach, Florida

Susan Gorman
Science Teacher
Northridge Middle School
North Richland Hills, Texas

C. John Graves
Science Teacher
Monforton Middle School
Bozeman, Montana

Janel Guse
Science Teacher and Dept. Chair
West Central Middle School
Hartford, South Dakota

Gary Habeeb
Science Mentor
Sierra–Plumas Joint Unified
School District
Downieville, California

Dennis Hanson
Science Teacher and Dept. Chair
Big Bear Middle School
Big Bear Lake, California

Norman E. Holcomb
Science Teacher
Marion Local Schools
Maria Stein, Ohio

Tracy Jahn
Science Teacher
Berkshire Junior-Senior High
School
Canaan, New York

David D. Jones
Science Teacher
Andrew Jackson Middle School
Cross Lanes, West Virginia

Howard A. Knodle
Science Teacher
Belvidere High School
Belvidere, Illinois

Michael E. Kral
Science Teacher
West Hardin Middle School
Cecilia, Kentucky

Kathy LaRoe
Science Teacher
East Valley Middle School
East Helena, Montana

Scott Mandel, Ph.D.
*Director and Educational
Consultant*
Teachers Helping Teachers
Los Angeles, California

Kathy McKee
Science Teacher
Hoyt Middle School
Des Moines, Iowa

Michael Minium
*Vice President of Program
Development*
United States Orienteering
Federation
Forest Park, Georgia

Jan Nelson
Science Teacher
East Valley Middle School
East Helena, Montana

Dwight C. Patton
Science Teacher
Carroll T. Welch Middle School
Horizon City, Texas

Joseph Price
Chairman—Science Department
H. M. Brown Junior High
School
Washington, D.C.

Terry J. Rakes
Science Teacher
Elmwood Junior High School
Rogers, Arkansas

Steven Ramig
Science Teacher
West Point High School
West Point, Nebraska

Helen P. Schiller
Science Teacher
Northwood Middle School
Taylors, South Carolina

Bert J. Sherwood
Science Teacher
Socorro Middle School
El Paso, Texas

Larry Tackett
Science Teacher and Dept. Chair
Andrew Jackson Middle School
Cross Lanes, West Virginia

Walter Woolbaugh
Science Teacher
Manhattan Junior High School
Manhattan, Montana

Alexis S. Wright
Middle School Science Coordinator
Rye Country Day School
Rye, New York

Gordon Zibelman
Science Teacher
Drexel Hill Middle School
Drexel Hill, Pennsylvania

Earth's Changing Surface

G

Skills Development

Process Skills

QuickLabs

Chapter Labs

Research and Critical Thinking Skills

Apply

Feature Articles

Science, Technology, and Society

Careers
Across the Sciences
Eye on the Environment

Connections

Mathematics

Program Scope and Sequence

Selecting the right books for your course is easy. Just review the topics presented in each book to determine the best match to your district curriculum.

	A MICROORGANISMS, FUNGI, AND PLANTS	**B** ANIMALS
CHAPTER 1	**It's Alive!! Or, Is It?** ❑ Characteristics of living things ❑ Homeostasis ❑ Heredity and DNA ❑ Producers, consumers, and decomposers ❑ Biomolecules	**Animals and Behavior** ❑ Characteristics of animals ❑ Classification of animals ❑ Animal behavior ❑ Hibernation and estivation ❑ The biological clock ❑ Animal communication ❑ Living in groups
CHAPTER 2	**Bacteria and Viruses** ❑ Binary fission ❑ Characteristics of bacteria ❑ Nitrogen-fixing bacteria ❑ Antibiotics ❑ Pathogenic bacteria ❑ Characteristics of viruses ❑ Lytic cycle	**Invertebrates** ❑ General characteristics of invertebrates ❑ Types of symmetry ❑ Characteristics of sponges, cnidarians, arthropods, and echinoderms ❑ Flatworms versus roundworms ❑ Types of circulatory systems
CHAPTER 3	**Protists and Fungi** ❑ Characteristics of protists ❑ Types of algae ❑ Types of protozoa ❑ Protist reproduction ❑ Characteristics of fungi and lichens	**Fishes, Amphibians, and Reptiles** ❑ Characteristics of vertebrates ❑ Structure and kinds of fishes ❑ Development of lungs ❑ Structure and kinds of amphibians and reptiles ❑ Function of the amniotic egg
CHAPTER 4	**Introduction to Plants** ❑ Characteristics of plants and seeds ❑ Reproduction and classification ❑ Angiosperms versus gymnosperms ❑ Monocots versus dicots ❑ Structure and functions of roots, stems, leaves, and flowers	**Birds and Mammals** ❑ Structure and kinds of birds ❑ Types of feathers ❑ Adaptations for flight ❑ Structure and kinds of mammals ❑ Function of the placenta
CHAPTER 5	**Plant Processes** ❑ Pollination and fertilization ❑ Dormancy ❑ Photosynthesis ❑ Plant tropisms ❑ Seasonal responses of plants	
CHAPTER 6		
CHAPTER 7		

Life Science

C CELLS, HEREDITY, & CLASSIFICATION

Cells: The Basic Units of Life
- ❏ Cells, tissues, and organs
- ❏ Populations, communities, and ecosystems
- ❏ Cell theory
- ❏ Surface-to-volume ratio
- ❏ Prokaryotic versus eukaryotic cells
- ❏ Cell organelles

The Cell in Action
- ❏ Diffusion and osmosis
- ❏ Passive versus active transport
- ❏ Endocytosis versus exocytosis
- ❏ Photosynthesis
- ❏ Cellular respiration and fermentation
- ❏ Cell cycle

Heredity
- ❏ Dominant versus recessive traits
- ❏ Genes and alleles
- ❏ Genotype, phenotype, the Punnett square and probability
- ❏ Meiosis
- ❏ Determination of sex

Genes and Gene Technology
- ❏ Structure of DNA
- ❏ Protein synthesis
- ❏ Mutations
- ❏ Heredity disorders and genetic counseling

The Evolution of Living Things
- ❏ Adaptations and species
- ❏ Evidence for evolution
- ❏ Darwin's work and natural selection
- ❏ Formation of new species

The History of Life on Earth
- ❏ Geologic time scale and extinctions
- ❏ Plate tectonics
- ❏ Human evolution

Classification
- ❏ Levels of classification
- ❏ Cladistic diagrams
- ❏ Dichotomous keys
- ❏ Characteristics of the six kingdoms

D HUMAN BODY SYSTEMS & HEALTH

Body Organization and Structure
- ❏ Homeostasis
- ❏ Types of tissue
- ❏ Organ systems
- ❏ Structure and function of the skeletal system, muscular system, and integumentary system

Circulation and Respiration
- ❏ Structure and function of the cardiovascular system, lymphatic system, and respiratory system
- ❏ Respiratory disorders

The Digestive and Urinary Systems
- ❏ Structure and function of the digestive system
- ❏ Structure and function of the urinary system

Communication and Control
- ❏ Structure and function of the nervous system and endocrine system
- ❏ The senses
- ❏ Structure and function of the eye and ear

Reproduction and Development
- ❏ Asexual versus sexual reproduction
- ❏ Internal versus external fertilization
- ❏ Structure and function of the human male and female reproductive systems
- ❏ Fertilization, placental development, and embryo growth
- ❏ Stages of human life

Body Defenses and Disease
- ❏ Types of diseases
- ❏ Vaccines and immunity
- ❏ Structure and function of the immune system
- ❏ Autoimmune diseases, cancer, and AIDS

Staying Healthy
- ❏ Nutrition and reading food labels
- ❏ Alcohol and drug effects on the body
- ❏ Hygiene, exercise, and first aid

E ENVIRONMENTAL SCIENCE

Interactions of Living Things
- ❏ Biotic versus abiotic parts of the environment
- ❏ Producers, consumers, and decomposers
- ❏ Food chains and food webs
- ❏ Factors limiting population growth
- ❏ Predator-prey relationships
- ❏ Symbiosis and coevolution

Cycles in Nature
- ❏ Water cycle
- ❏ Carbon cycle
- ❏ Nitrogen cycle
- ❏ Ecological succession

The Earth's Ecosystems
- ❏ Kinds of land and water biomes
- ❏ Marine ecosystems
- ❏ Freshwater ecosystems

Environmental Problems and Solutions
- ❏ Types of pollutants
- ❏ Types of resources
- ❏ Conservation practices
- ❏ Species protection

Energy Resources
- ❏ Types of resources
- ❏ Energy resources and pollution
- ❏ Alternative energy resources

Scope and Sequence *(continued)*

Earth Science

H WATER ON EARTH

The Flow of Fresh Water
- ❏ Water cycle
- ❏ River systems
- ❏ Stream erosion
- ❏ Life cycle of rivers
- ❏ Deposition
- ❏ Aquifers, springs, and wells
- ❏ Ground water
- ❏ Water treatment and pollution

Exploring the Oceans
- ❏ Properties and characteristics of the oceans
- ❏ Features of the ocean floor
- ❏ Ocean ecology
- ❏ Ocean resources and pollution

The Movement of Ocean Water
- ❏ Types of currents
- ❏ Characteristics of waves
- ❏ Types of ocean waves
- ❏ Tides

I WEATHER AND CLIMATE

The Atmosphere
- ❏ Structure of the atmosphere
- ❏ Air pressure
- ❏ Radiation, convection, and conduction
- ❏ Greenhouse effect and global warming
- ❏ Characteristics of winds
- ❏ Types of winds
- ❏ Air pollution

Understanding Weather
- ❏ Water cycle
- ❏ Humidity
- ❏ Types of clouds
- ❏ Types of precipitation
- ❏ Air masses and fronts
- ❏ Storms, tornadoes, and hurricanes
- ❏ Weather forecasting
- ❏ Weather maps

Climate
- ❏ Weather versus climate
- ❏ Seasons and latitude
- ❏ Prevailing winds
- ❏ Earth's biomes
- ❏ Earth's climate zones
- ❏ Ice ages
- ❏ Global warming
- ❏ Greenhouse effect

J ASTRONOMY

Observing the Sky
- ❏ Astronomy
- ❏ Keeping time
- ❏ Mapping the stars
- ❏ Scales of the universe
- ❏ Types of telescope
- ❏ Radioastronomy

Formation of the Solar System
- ❏ Birth of the solar system
- ❏ Planetary motion
- ❏ Newton's Law of Universal Gravitation
- ❏ Structure of the sun
- ❏ Fusion
- ❏ Earth's structure and atmosphere

A Family of Planets
- ❏ Properties and characteristics of the planets
- ❏ Properties and characteristics of moons
- ❏ Comets, asteroids, and meteoroids

The Universe Beyond
- ❏ Composition of stars
- ❏ Classification of stars
- ❏ Star brightness, distance, and motions
- ❏ H-R diagram
- ❏ Life cycle of stars
- ❏ Types of galaxies
- ❏ Theories on the formation of the universe

Exploring Space
- ❏ Rocketry and artificial satellites
- ❏ Types of Earth orbit
- ❏ Space probes and space exploration

Scope and Sequence (continued)

	K INTRODUCTION TO MATTER	L INTERACTIONS OF MATTER
CHAPTER 1	**The Properties of Matter** ❏ Definition of matter ❏ Mass and weight ❏ Physical and chemical properties ❏ Physical and chemical change ❏ Density	**Chemical Bonding** ❏ Types of chemical bonds ❏ Valence electrons ❏ Ions versus molecules ❏ Crystal lattice
CHAPTER 2	**States of Matter** ❏ States of matter and their properties ❏ Boyle's and Charles's laws ❏ Changes of state	**Chemical Reactions** ❏ Writing chemical formulas and equations ❏ Law of conservation of mass ❏ Types of reactions ❏ Endothermic versus exothermic reactions ❏ Law of conservation of energy ❏ Activation energy ❏ Catalysts and inhibitors
CHAPTER 3	**Elements, Compounds, and Mixtures** ❏ Elements and compounds ❏ Metals, nonmetals, and metalloids (semiconductors) ❏ Properties of mixtures ❏ Properties of solutions, suspensions, and colloids	**Chemical Compounds** ❏ Ionic versus covalent compounds ❏ Acids, bases, and salts ❏ pH ❏ Organic compounds ❏ Biomolecules
CHAPTER 4	**Introduction to Atoms** ❏ Atomic theory ❏ Atomic model and structure ❏ Isotopes ❏ Atomic mass and mass number	**Atomic Energy** ❏ Properties of radioactive substances ❏ Types of decay ❏ Half-life ❏ Fission, fusion, and chain reactions
CHAPTER 5	**The Periodic Table** ❏ Structure of the periodic table ❏ Periodic law ❏ Properties of alkali metals, alkaline-earth metals, halogens, and noble gases	
CHAPTER 6		

Physical Science

HOLT SCIENCE & TECHNOLOGY

Components Listing

Effective planning starts with all the resources you need in an easy-to-use package for each short course.

Directed Reading Worksheets Help students develop and practice fundamental reading comprehension skills and provide a comprehensive review tool for students to use when studying for an exam.

Study Guide Vocabulary & Notes Worksheets and Chapter Review Worksheets are reproductions of the Chapter Highlights and Chapter Review sections that follow each chapter in the textbook.

Science Puzzlers, Twisters & Teasers Use vocabulary and concepts from each chapter of the Pupil's Editions as elements of rebuses, anagrams, logic puzzles, daffy definitions, riddle poems, word jumbles, and other types of puzzles.

Reinforcement and Vocabulary Review Worksheets Approach a chapter topic from a different angle with an emphasis on different learning modalities to help students that are frustrated by traditional methods.

Critical Thinking & Problem Solving Worksheets Develop the following skills: distinguishing fact from opinion, predicting consequences, analyzing information, and drawing conclusions. Problem Solving Worksheets develop a step-by-step process of problem analysis including gathering information, asking critical questions, identifying alternatives, and making comparisons.

Math Skills for Science Worksheets Each activity gives a brief introduction to a relevant math skill, a step-by-step explanation of the math process, one or more example problems, and a variety of practice problems.

Science Skills Worksheets Help your students focus specifically on skills such as measuring, graphing, using logic, understanding statistics, organizing research papers, and critical thinking options.

LAB ACTIVITIES

Datasheets for Labs These worksheets are the labs found in the *Holt Science & Technology* textbook. Charts, tables, and graphs are included to make data collection and analysis easier, and space is provided to write observations and conclusions.

Whiz-Bang Demonstrations Discovery or Making Models experiences label each demo as one in which students discover an answer or use a scientific model.

Calculator-Based Labs Give students the opportunity to use graphing-calculator probes and sensors to collect data using a TI graphing calculator, Vernier sensors, and a TI CBL 2™ or Vernier Lab Pro interface.

EcoLabs and Field Activities Focus on educational outdoor projects, such as wildlife observation, nature surveys, or natural history.

Inquiry Labs Use the scientific method to help students find their own path in solving a real-world problem.

Long-Term Projects and Research Ideas Provide students with the opportunity to go beyond library and Internet resources to explore science topics.

ASSESSMENT

Chapter Tests Each four-page chapter test consists of a variety of item types including Multiple Choice, Using Vocabulary, Short Answer, Critical Thinking, Math in Science, Interpreting Graphics, and Concept Mapping.

Performance-Based Assessments Evaluate students' abilities to solve problems using the tools, equipment, and techniques of science. Rubrics included for each assessment make it easy to evaluate student performance.

TEACHER RESOURCES

Lesson Plans Integrate all of the great resources in the *Holt Science & Technology* program into your daily teaching. Each lesson plan includes a correlation of the lesson activities to the National Science Education Standards.

Teaching Transparencies Each transparency is correlated to a particular lesson in the Chapter Organizer.

Concept Mapping Transparencies, Worksheets, and Answer Key

Give students an opportunity to complete their own concept maps to study the concepts within each chapter and form logical connections. Student worksheets contain a blank concept map with linking phrases and a list of terms to be used by the student to complete the map.

TECHNOLOGY RESOURCES

 One-Stop Planner CD-ROM

Finding the right resources is easy with the One-Stop Planner CD-ROM. You can view and print any resource with just the click of a mouse. Customize the suggested lesson plans to match your daily or weekly calendar and your district's requirements. Powerful test generator software allows you to create customized assessments using a databank of items.

The One-Stop Planner for each level includes the following:

- All materials from the Teaching Resources
- Bellringer Transparency Masters
- Block Scheduling Tools
- Standards Correlations
- Lab Inventory Checklist
- Safety Information
- Science Fair Guide
- Parent Involvement Tools
- Spanish Audio Scripts
- Spanish Glossary
- Assessment Item Listing
- Assessment Checklists and Rubrics
- Test Generator

 sciLINKS

sciLINKS numbers throughout the text take you and your students to some of the best on-line resources available. Sites are constantly reviewed and updated by the National Science Teachers Association. Special "teacher only" sites are available to you once you register with the service.

 go.hrw.com

To access Holt, Rinehart and Winston Web resources, use the home page codes for each level found on page 1 of the Pupil's Editions. The codes shown on the Chapter Organizers for each chapter in the Annotated Teacher's Edition take you to chapter-specific resources.

 Smithsonian Institution

Find lesson plans, activities, interviews, virtual exhibits, and just general information on a wide variety of topics relevant to middle school science.

CNNfyi.com

Find the latest in late-breaking science news for students. Featured news stories are supported with lesson plans and activities.

 Presents Science in the News Video Library

Bring relevant science news stories into the classroom. Each video comes with a Teacher's Guide and set of Critical Thinking Worksheets that develop listening and media analysis skills. Tapes in the series include:

- Eye on the Environment
- Multicultural Connections
- Scientists in Action
- Science, Technology & Society

 Guided Reading Audio CD Program

Students can listen to a direct read of each chapter and follow along in the text. Use the program as a content bridge for struggling readers and students for whom English is not their native language.

 Interactive Explorations CD-ROM

Turn a computer into a virtual laboratory. Students act as lab assistants helping Dr. Crystal Labcoat solve real-world problems. Activities develop students' inquiry, analysis, and decision-making skills.

Interactive Science Encyclopedia CD-ROM

Give your students access to more than 3,000 cross-referenced scientific definitions, in-depth articles, science fair project ideas, activities, and more.

ADDITIONAL COMPONENTS

Holt Anthology of Science Fiction

Science Fiction features in the Pupil's Edition preview the stories found in the anthology. Each story begins with a Reading Prep guide and closes with Think About It questions.

Professional Reference for Teachers

Articles written by leading educators help you learn more about the National Science Education Standards, block scheduling, classroom management techniques, and more. A bibliography of professional references is included.

Holt Science Posters

Seven wall posters highlight interesting topics, such as the Physics of Sports, or useful reference material, such as the Scientific Method.

 Holt Science Skills Workshop: Reading in the Content Area

Use a variety of in-depth skills exercises to help students learn to read science materials strategically.

Key

 These materials are blackline masters.

 All titles shown in green are found in the *Teaching Resources* booklets for each course.

Science & Math Skills Worksheets

The *Holt Science and Technology* program helps you meet the needs of a wide variety of students, regardless of their skill level. The following pages provide examples of the worksheets available to improve your students' science and math skills, whether they already have a strong science and math background or are weak in these areas. Samples of assessment checklists and rubrics are also provided.

In addition to the skills worksheets represented here, *Holt Science and Technology* provides a variety of worksheets that are correlated directly with each chapter of the program. Representations of these worksheets are found at the beginning of each chapter in this Annotated Teacher's Edition. Specific worksheets related to each chapter are listed in the Chapter Organizer. Worksheets and transparencies are found in the softcover *Teaching Resources* for each course.

Many worksheets are also available on the HRW Web site. The address is **go.hrw.com.**

Science Skills Worksheets: Thinking Skills

BEING FLEXIBLE

USING YOUR SENSES
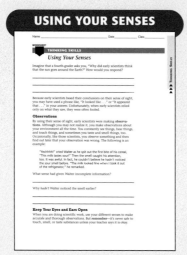

THINKING OBJECTIVELY
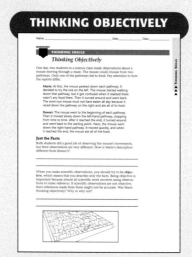

UNDERSTANDING BIAS
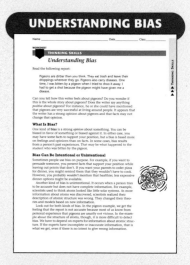

USING LOGIC

BOOSTING YOUR MEMORY

IMPROVING YOUR STUDY HABITS

READING A SCIENCE TEXTBOOK
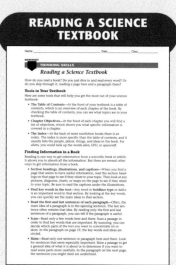

Science Skills Worksheets: Experimenting Skills

SAFETY RULES!
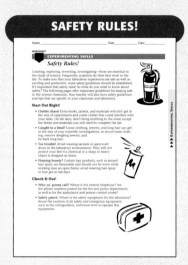

DOING A LAB WRITE-UP
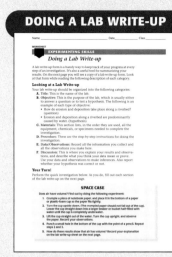

UNDERSTANDING VARIABLES

WORKING WITH HYPOTHESES
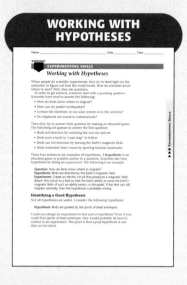

DESIGNING AN EXPERIMENT

USING THE INTERNATIONAL SYSTEM OF UNITS (SI)
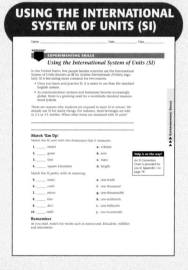

MEASURING
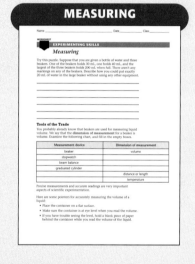

Science Skills Worksheets: Researching Skills

CHOOSING YOUR TOPIC

ORGANIZING YOUR RESEARCH
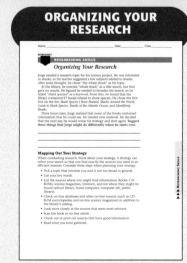

FINDING USEFUL SOURCES
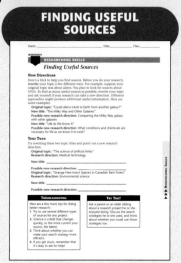

RESEARCHING ON THE WEB

Science Skills Worksheets: Researching Skills (continued)

IDENTIFYING BIAS

TAKING NOTES

SCIENCE WRITING

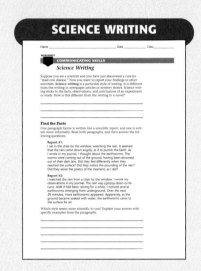

Science Skills Worksheets: Communicating Skills

SCIENCE DRAWING

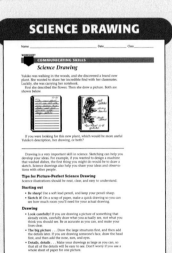

USING MODELS TO COMMUNICATE

INTRODUCTION TO GRAPHS

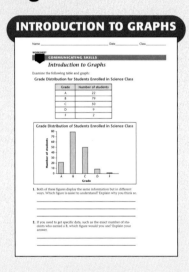

GRASPING GRAPHING

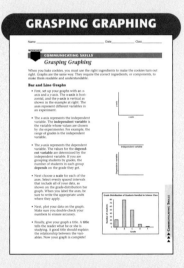

INTERPRETING YOUR DATA

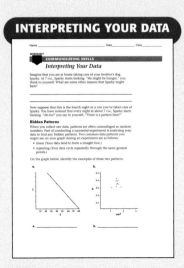

RECOGNIZING BIAS IN GRAPHS

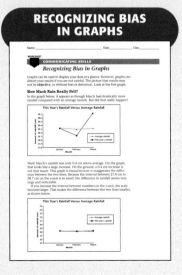

MAKING DATA MEANINGFUL

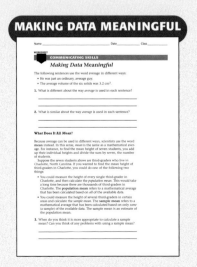

HINTS FOR ORAL PRESENTATIONS

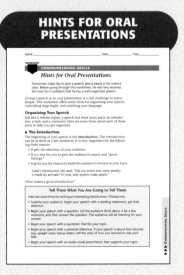

Math Skills for Science

ADDITION AND SUBTRACTION

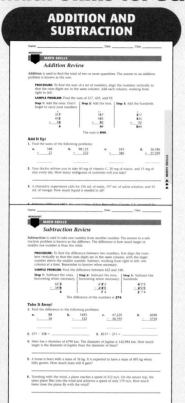

MULTIPLICATION

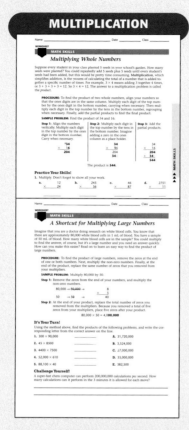

DIVISION

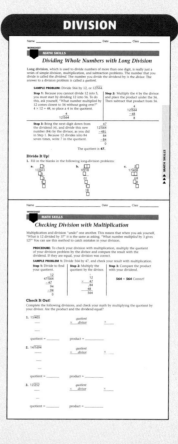

AVERAGES

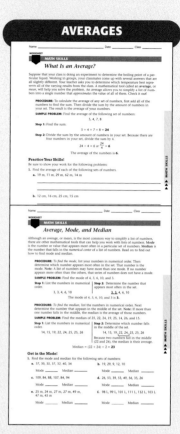

POSITIVE AND NEGATIVE NUMBERS

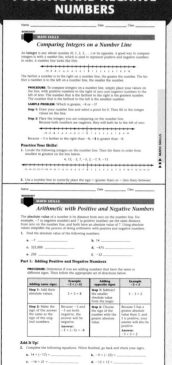

FRACTIONS

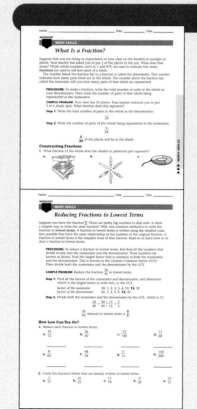

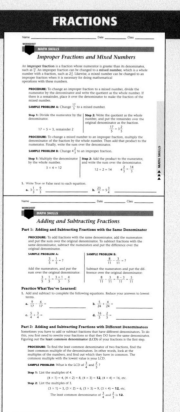

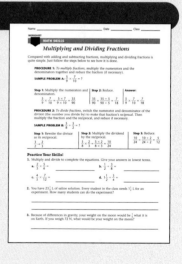

Math Skills for Science (continued)

RATIOS AND PROPORTIONS
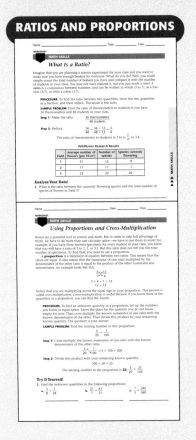

DECIMALS
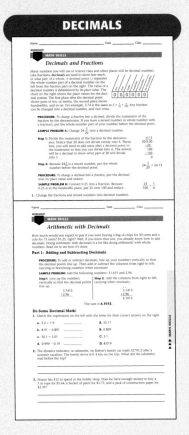

PERCENTAGES
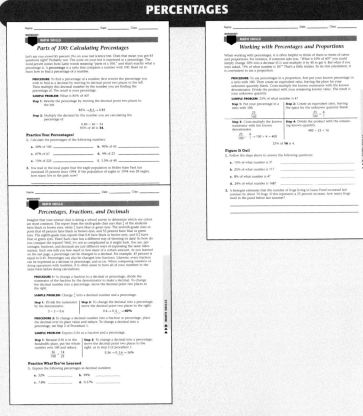

POWERS OF 10
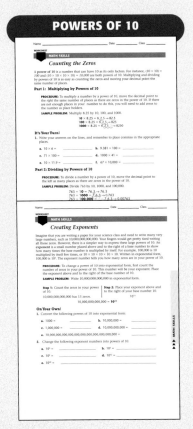

SCIENTIFIC NOTATION
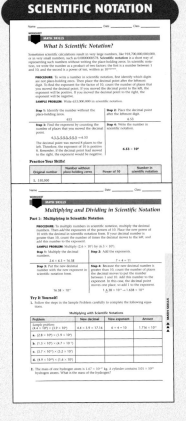

SI MEASUREMENT AND CONVERSION
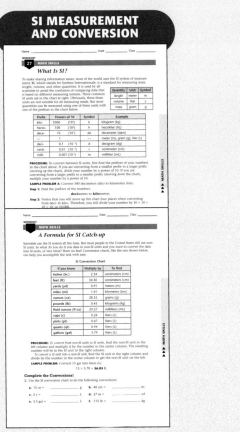

Math Skills for Science (continued)

GEOMETRY

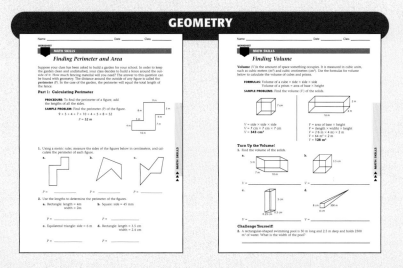

THE UNIT FACTOR AND DIMENSIONAL ANALYSIS

MATH IN SCIENCE: INTEGRATED SCIENCE

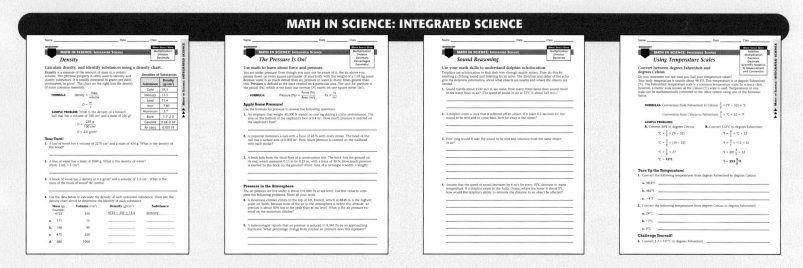

Science & Math Skills Worksheets (continued)

Math Skills for Science (continued)

MATH IN SCIENCE: EARTH SCIENCE

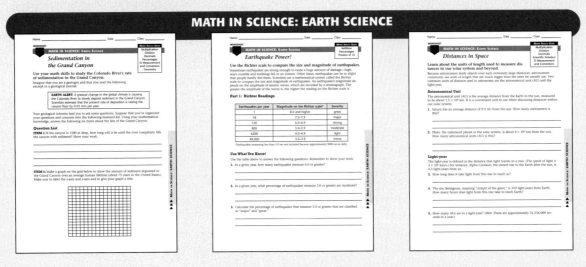

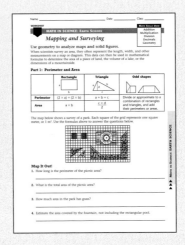

Assessment Checklist & Rubrics

The following is just a sample of over 50 checklists and rubrics contained in this booklet.

RUBRICS FOR WRITTEN WORK

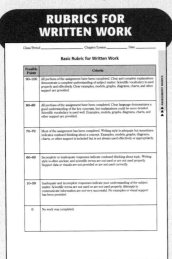

RUBRIC FOR EXPERIMENTS

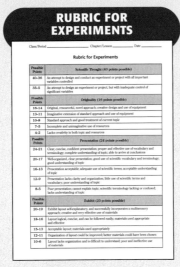

TEACHER EVALUATION OF COOPERATIVE LEARNING

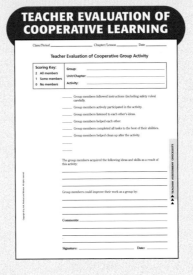

TEACHER EVALUATION OF STUDENT PROGRESS

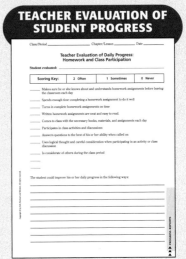

EARTH SCIENCE NATIONAL SCIENCE EDUCATION STANDARDS CORRELATIONS

The following lists show the chapter correlation of **Holt Science and Technology: Earth's Changing Surface** with the *National Science Education Standards* (grades 5-8)

UNIFYING CONCEPTS AND PROCESSES

Standard	Chapter Correlation	
Evidence, models, and explanation Code: UCP 2	Chapter 1	1.1, 1.2, 1.3
	Chapter 2	2.1
	Chapter 3	3.2, 3.3
Change, constancy, and measurement Code: UCP 3	Chapter 1	1.1, 1.2, 1.3
	Chapter 3	3.1, 3.2, 3.3
Form and function Code: UCP 5	Chapter 1	1.1, 1.2, 1.3

SCIENCE IN PERSONAL AND SOCIAL PERSPECTIVES

Standard	Chapter Correlation	
Populations, resources, and environments Code: SPSP 2	Chapter 1	1.3
	Chapter 2	2.4
	Chapter 3	3.1, 3.2
Natural hazards Code: SPSP 3	Chapter 3	3.1, 3.3, 3.4
Risks and benefits Code: SPSP 4	Chapter 2	2.4
	Chapter 3	3.4
Science and technology in society Code: SPSP 5	Chapter 1	1.1, 1.2, 1.3
	Chapter 2	2.4

SCIENCE AS INQUIRY

Standard	Chapter Correlation	
Abilities necessary to do scientific inquiry Code: SAI 1	Chapter 1	1.1, 1.2, 1.3
	Chapter 2	2.1, 2.2
	Chapter 3	3.1, 3.2, 3.3, 3.4

SCIENCE AND TECHNOLOGY

Standard	Chapter Correlation	
Abilities of technological design Code: ST 1	Chapter 1	1.1, 1.3
Understandings about science and technology Code: ST 2	Chapter 1	1.1, 1.2, 1.3
	Chapter 3	3.2

HISTORY AND NATURE OF SCIENCE

Standard	Chapter Correlation	
Science as a human endeavor Code: HNS 1	Chapter 1	1.1, 1.3
	Chapter 3	3.2
Nature of science Code: HNS 2	Chapter 3	3.2
History of science Code: HNS 3	Chapter 1	1.1, 1.2
	Chapter 3	3.2

EARTH SCIENCE NATIONAL SCIENCE EDUCATION CONTENT STANDARDS

STRUCTURE OF THE EARTH SYSTEM	
Standard	**Chapter Correlation**
Land forms are the result of a combination of constructive and destructive forces. Constructive forces include crustal deformation, volcanic eruption, and deposition of sediment, while destructive forces include weathering and erosion. Code: ES 1c	**Chapter 2** 2.1, 2.2, 2.3 **Chapter 3** 3.1, 3.2, 3.3, 3.4
Living organisms have played many roles in the earth system, including affecting the composition of the atmosphere, producing some types of rocks, and contributing to the weathering of rocks. Code: ES 1k	**Chapter 2** 2.1, 2.3

EARTH'S HISTORY	
Standard	**Chapter Correlation**
The earth processes we see today, including erosion, movement of lithospheric plates, and changes in atmospheric composition, are similar to those that occurred in the past. Earth history is also influenced by occasional catastrophes, such as the impact of an asteroid or comet. Code: ES 2a	**Chapter 3** 3.2, 3.3

Master Materials List

For added convenience, Science Kit® provides materials-ordering software on CD-ROM designed specifically for *Holt Science and Technology*. Using this software, you can order complete kits or individual items, quickly and efficiently.

CONSUMABLE MATERIALS	AMOUNT	PAGE
Bag, plastic sealable sandwich	3	94
Bottle, soda, 3 L (or jar)	1	46
Clay, modeling	1 stick	20
Cup, clear plastic, 9 fl oz	2	29
Dust	1 lb	63
Ice, cubed	4–5	79
Ketchup	25 mL	34
Limestone	24 pieces	46
Marker, permanent, black	1	46, 78
Marker, transparency	1	92
Mask, disposable filter	1	63, 78
Paper, graph	1 sheet	46
Paper, tracing	1 sheet	92
Paper, white	1 sheet	3
Pencil, assorted, colored	1 box	3

CONSUMABLE EQUIPMENT	AMOUNT	PAGE
Pencil, colored (or marker)	2	90
Poster board	1 sheet	46
Potato	1	92
Rubber band	3–4	94
Sand	800 mL	74
Sand	10 lb	79
Sand	15 lb	55
Sand, fine	1 lb	63, 78, 95
Spoon, plastic	2	29
String	1 ball	20, 94
Sugar cube	1	29
Sugar, granulated	1 tsp	29
Tape, masking	1 roll	20
Tape, transparent	1 roll	94

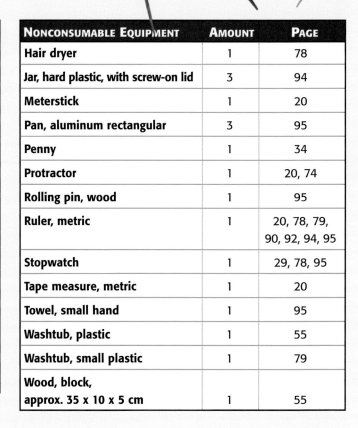

Nonconsumable Equipment	Amount	Page
Bag, paper, large	1	78
Basketball	1	20
Box, cardboard, shallow (or lid)	1	78
Brick	3	95
Calculator, scientific	1	20
Clay, modeling	2 lb	95
Compass, magnetic	1	6, 90
Container, clear plastic storage with lid	1	92
Container, empty large margarine	3	95
Fan, electric	1	63
Flashlight	1	20
Goggles, safety	1	63
Graduated cylinder, 50 mL	1	95
Gravel	1 lb	63, 95

Nonconsumable Equipment	Amount	Page
Hair dryer	1	78
Jar, hard plastic, with screw-on lid	3	94
Meterstick	1	20
Pan, aluminum rectangular	3	95
Penny	1	34
Protractor	1	20, 74
Rolling pin, wood	1	95
Ruler, metric	1	20, 78, 79, 90, 92, 94, 95
Stopwatch	1	29, 78, 95
Tape measure, metric	1	20
Towel, small hand	1	95
Washtub, plastic	1	55
Washtub, small plastic	1	79
Wood, block, approx. 35 x 10 x 5 cm	1	55

Answers to Concept Mapping Questions

The following pages contain sample answers to all of the concept mapping questions that appear in the Chapter Reviews. Because there is more than one way to do a concept map, your students' answers may vary.

CHAPTER 1 Maps as Models of the Earth

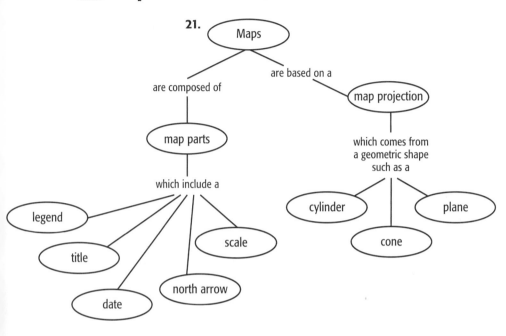

CHAPTER 2 Weathering and Soil Formation

CHAPTER 3 Agents of Erosion and Deposition

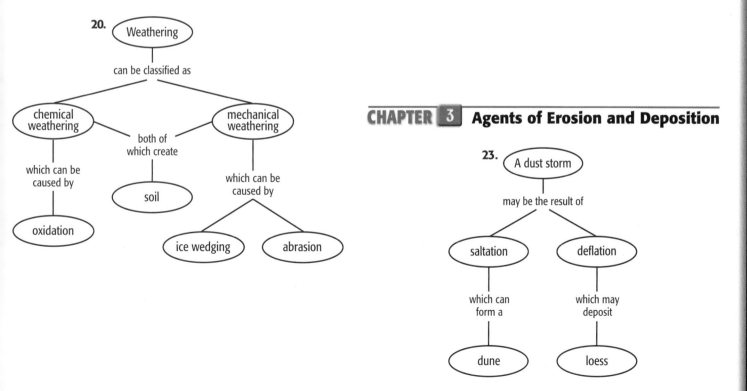

To the Student

This book was created to make your science experience interesting, exciting, and fun!

Go for It!

Science is a process of discovery, a trek into the unknown. The skills you develop using *Holt Science & Technology*— such as observing, experimenting, and explaining observations and ideas— are the skills you will need for the future. There is a universe of exploration and discovery awaiting those who accept the challenges of science.

Science & Technology

You see the interaction between science and technology every day. Science makes technology possible. On the other hand, some of the products of technology, such as computers, are used to make further scientific discoveries. In fact, much of the scientific work that is done today has become so technically complicated and expensive that no one person can do it entirely alone. But make no mistake, the creative ideas for even the most highly technical and expensive scientific work still come from individuals.

Activities and Labs

The activities and labs in this book will allow you to make some basic but important scientific discoveries on your own. You can even do some exploring on your own at home! Here's your chance to use your imagination and curiosity as you investigate your world.

Keep a ScienceLog

In this book, you will be asked to keep a type of journal called a ScienceLog to record your thoughts, observations, experiments, and conclusions. As you develop your ScienceLog, you will see your own ideas taking shape over time. You'll have a written record of how your ideas have changed as you learn about and explore interesting topics in science.

Know "What You'll Do"

The "What You'll Do" list at the beginning of each section is your built-in guide to what you need to learn in each chapter. When you can answer the questions in the Section Review and Chapter Review, you know you are ready for a test.

Check Out the Internet

You will see this  logo throughout the book. You'll be using *sci*LINKS as your gateway to the Internet. Once you log on to *sci*LINKS using your computer's Internet link, type in the *sci*LINKS address. When asked for the keyword code, type in the keyword for that topic. A wealth of resources is now at your disposal to help you learn more about that topic.

In addition to *sci*LINKS you can log on to some other great resources to go with your text. The addresses shown below will take you to the home page of each site.

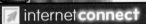

This textbook contains the following on-line resources to help you make the most of your science experience.

Visit **go.hrw.com** for extra help and study aids matched to your textbook. Just type in the keyword HR2 HOME.

Visit **www.scilinks.org** to find resources specific to topics in your textbook. Keywords appear throughout your book to take you further.

 Smithsonian Institution®
Internet Connections

Visit **www.si.edu/hrw** for specifically chosen on-line materials from one of our nation's premier science museums.

Visit **www.cnnfyi.com** for late-breaking news and current events stories selected just for you.

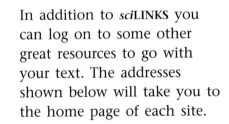

Chapter Organizer

CHAPTER ORGANIZATION	TIME MINUTES	OBJECTIVES	LABS, INVESTIGATIONS, AND DEMONSTRATIONS
Chapter Opener pp. 2–3	45	National Standards: SAI 1, SPSP 5, HNS 1, 3	**Start-Up Activity,** Follow the Yellow Brick Road, p. 3
Section 1 You Are Here	90	▶ Explain how a magnetic compass can be used to find directions on the Earth. ▶ Distinguish between true north and magnetic north. ▶ Distinguish between lines of latitude and lines of longitude on a globe or map. ▶ Explain how latitude and longitude can be used to locate places on Earth. UCP 2, 3, 5, SAI 1, ST 2, SPSP 5, HNS 1, 3; Labs UCP 2, 3, SAI 1, ST 1, HNS 1	**QuickLab,** Finding Directions with a Compass, p. 6 **Skill Builder,** Round or Flat? p. 20 **Datasheets for LabBook,** Round or Flat? **Design Your Own,** Orient Yourself! p. 90 **Datasheets for LabBook,** Orient Yourself!
Section 2 Mapping the Earth's Surface	90	▶ Compare a map with a globe. ▶ Describe the three types of map projections. ▶ Describe recent technological advances that have helped the science of mapmaking progress. ▶ List the parts of a map. UCP 2, 3, 5, SAI 1, ST 2, SPSP 5, HNS 3	
Section 3 Topographic Maps	135	▶ Describe how contour lines show elevation and landforms on a map. ▶ List the rules of contour lines. ▶ Interpret a topographic map. UCP 2, 3, 5, SAI 1, ST 2, SPSP 2, 5, HNS 1; Labs UCP 2, 3, SAI 1, ST 1	**Skill Builder,** Topographic Tuber, p. 92 **Datasheets for LabBook,** Topographic Tuber **Inquiry Labs,** Looking for Buried Treasure **Long-Term Projects & Research Ideas,** Globe Trotting

*See page **T23** for a complete correlation of this book with the*

NATIONAL SCIENCE EDUCATION STANDARDS.

TECHNOLOGY RESOURCES

 Guided Reading Audio CD
English or Spanish, Chapter 1

 Science Discovery Videodiscs
Image and Activity Bank with Lesson Plans: Remote Sensing

 CNN. Multicultural Connections, Mapping Asian Temples from Space, Segment 2

 One-Stop Planner CD-ROM with Test Generator

Chapter 1 • Maps as Models of the Earth

CLASSROOM WORKSHEETS, TRANSPARENCIES, AND RESOURCES	SCIENCE INTEGRATION AND CONNECTIONS	REVIEW AND ASSESSMENT
Directed Reading Worksheet **Science Puzzlers, Twisters & Teasers**	**Cross-Disciplinary Focus,** p. 3 in ATE	
Transparency 104, Finding Direction on Earth **Directed Reading Worksheet,** Section 1 **Transparency 105,** Lines of Latitude **Transparency 105,** Lines of Longitude **Reinforcement Worksheet,** Where on Earth?	**Multicultural Connection,** p. 4 in ATE **Connect to Life Science,** p. 5 in ATE **Multicultural Connection,** p. 7 in ATE **Astronomy Connection,** p. 7	**Self-Check,** p. 6 **Homework,** p. 8 in ATE **Section Review,** p. 9 **Quiz,** p. 9 in ATE **Alternative Assessment,** p. 9 in ATE
Directed Reading Worksheet, Section 2 **Transparency 106,** Mercator Projection **Transparency 106,** Conic Projection **Transparency 106,** Azimuthal Projection **Transparency 291,** The Electromagnetic Spectrum **Math Skills for Science Worksheet,** Using Proportions and Cross-Multiplication **Critical Thinking Worksheet,** Shaping the World	**Cross-Disciplinary Focus,** p. 12 in ATE **Cross-Disciplinary Focus,** p. 13 in ATE **Connect to Physical Science,** p. 13 in ATE **Math and More,** p. 14 in ATE **Multicultural Connection,** p. 14 in ATE **Apply,** p. 15 **Science, Technology, and Society:** The Lost City of Ubar, p. 26	**Homework,** p. 12 in ATE **Section Review,** p. 15 **Quiz,** p. 15 in ATE **Alternative Assessment,** p. 15 in ATE
Directed Reading Worksheet, Section 3 **Reinforcement Worksheet,** Interpreting a Topographic Map **Math Skills for Science Worksheet,** Mapping and Surveying	**MathBreak,** Counting Contours, p. 17 **Connect to Oceanography,** p. 18 in ATE **Environment Connection,** p. 19 **Careers:** Watershed Planner–Nancy Charbeneau, p. 27	**Self-Check,** p. 17 **Homework,** p. 17 in ATE **Section Review,** p. 19 **Quiz,** p. 19 in ATE **Alternative Assessment,** p. 19 in ATE

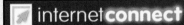 internet**connect**

 go.hrw.com **Holt, Rinehart and Winston On-line Resources**
go.hrw.com

For worksheets and other teaching aids related to this chapter, visit the HRW Web site and type in the keyword: **HSTMAP**

 SCI**LINKS** NSTA **National Science Teachers Association**
www.scilinks.org

Encourage students to use the *sci*LINKS numbers listed in the internet connect boxes to access information and resources on the **NSTA** Web site.

END-OF-CHAPTER REVIEW AND ASSESSMENT

Chapter Review in Study Guide
Vocabulary and Notes in Study Guide
Chapter Tests with Performance-Based Assessment, Chapter 1 Test
Chapter Tests with Performance-Based Assessment, Performance-Based Assessment 1
Concept Mapping Transparency 2

Chapter Resources & Worksheets

Visual Resources

TEACHING TRANSPARENCIES

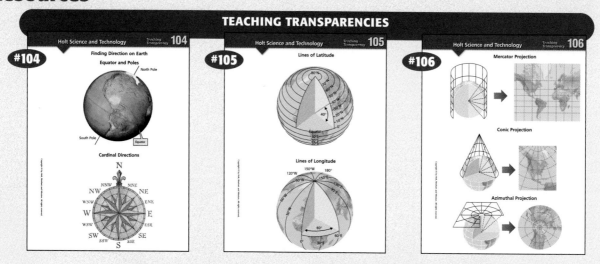

#104 — Finding Direction on Earth — Equator and Poles — Cardinal Directions

#105 — Lines of Latitude — Lines of Longitude

#106 — Mercator Projection — Conic Projection — Azimuthal Projection

TEACHING TRANSPARENCIES

CONCEPT MAPPING TRANSPARENCY

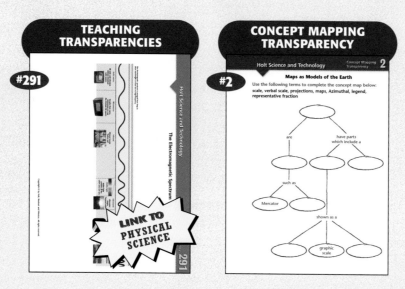

#291 — LINK TO PHYSICAL SCIENCE

#2 — Maps as Models of the Earth — Use the following terms to complete the concept map below: scale, verbal scale, projections, maps, Azimuthal, legend, representative fraction

Meeting Individual Needs

DIRECTED READING

#1 — DIRECTED READING WORKSHEET — Maps as Models of the Earth

REINFORCEMENT & VOCABULARY REVIEW

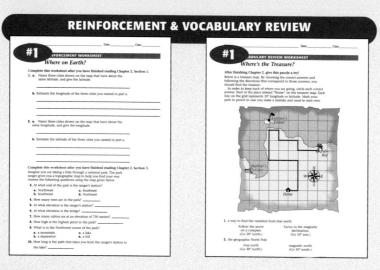

#1 — REINFORCEMENT WORKSHEET — Where on Earth?

#1 — VOCABULARY REVIEW WORKSHEET — Where's the Treasure?

SCIENCE PUZZLERS, TWISTERS & TEASERS

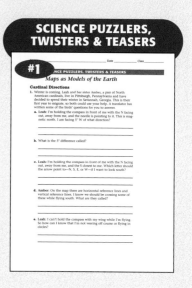

#1 — SCIENCE PUZZLERS, TWISTERS & TEASERS — Maps as Models of the Earth

Review & Assessment

STUDY GUIDE

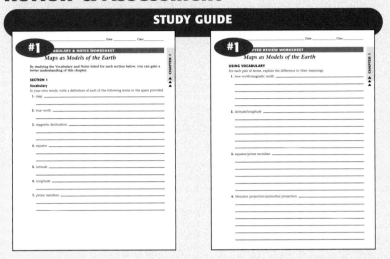

CHAPTER TESTS WITH PERFORMANCE-BASED ASSESSMENT

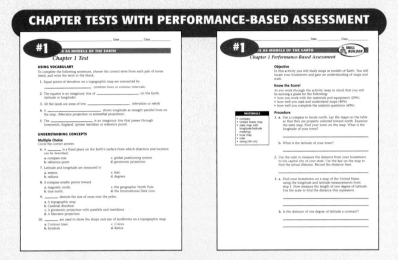

Lab Worksheets

INQUIRY LABS

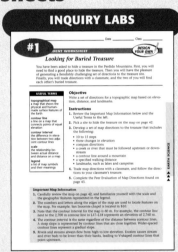

LONG-TERM PROJECTS & RESEARCH IDEAS

DATASHEETS FOR LABBOOK

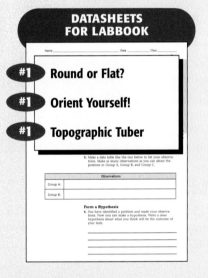

Applications & Extensions

CRITICAL THINKING & PROBLEM SOLVING

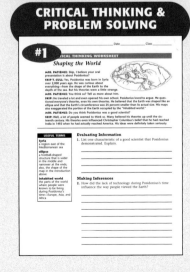

MULTICULTURAL CONNECTIONS

SECTION 1

You Are Here

▶ Global Positioning Systems

During the 1970s, the U.S. Department of Defense developed the Global Positioning System (GPS) for use in aircraft navigation and missile guidance. The system uses a network of 25 satellites that continually transmit positioning information to receivers on Earth. The distance between a receiver and at least three satellites is used to compute the latitude and longitude coordinates of the receiver's position.

- In 1983, the system was made available to the public and has since been used for land, sea, and air navigation, surveying, geophysical exploration, and vehicle-location systems. The system can be highly accurate, making it possible to determine position to within less than 1 m. As the prices of some of the receivers have plummeted, the use of the GPS for recreational activities, such as boating, hiking, and hunting, has increased.

IS THAT A FACT!

- ➡ The first GPS receiver was about the size of a filing cabinet, but by 1989, manufacturers had perfected hand-held versions.

▶ Longitude

Because Earth rotates 360° every 24 hours, it turns 15° every hour. It is therefore possible to determine longitude at any place on the globe if the local time and the time at the prime meridian are known. Before the mid-eighteenth century, however, the unreliability of clocks—especially those aboard ships, where motion, temperature variation, and moisture could wreak havoc with a timepiece's workings—thwarted calculations of longitude. Many shipwrecks were caused because the ship captains could not accurately calculate their location.

- In 1707, inaccurate longitudinal information caused four ships in a British fleet to run aground, and 2,000 sailors died. The British Parliament addressed the problem by offering a large reward to anyone who could develop a method to accurately calculate longitude within half a degree. John Harrison (1693–1776), a self-taught clockmaker, developed a chronometer that remained accurate on rough seas and won the prize in 1763. More than 200 years later, astronaut Neil Armstrong gave credit to Harrison for the role he played in enabling exploration of Earth and in inspiring future generations to venture toward exploration of the moon.

SECTION 2

Mapping the Earth's Surface

▶ Gerardus Mercator

Gerardus Mercator was born Gerhard Kremer in 1512, in Rupelmonde, Flanders (present-day Belgium). At age 24, Mercator was a highly skilled engraver, calligrapher, and scientific-instrument maker. With two of his teachers, he made the first globe of Earth, in 1536–1537. A true Renaissance man, Mercator was a highly esteemed cartographer who also published a treatise on italic lettering, designed a grammar-school curriculum, taught mathematics, and conducted genealogical research for his patron, Duke Wilhelm of Cleve. He even attempted to write a chronology of the history of the world from the formation of Earth to 1568.

▶ Aerial Photographs

Aerial photographs used for mapping have been taken from mountaintops, airplanes, hot-air balloons, rockets, and satellites. Originally, the photographs were taken from various angles. Today, specialized cameras take pictures straight down, in sequences, and along predetermined lines. Each picture overlaps parts of the pictures taken before and after it. This is done because the least amount of

distortion is at the center of each photo. Together these photos produce an image with very little distortion. When used with specialized instruments, these photos can be used to produce a three-dimensional view of an area. From the three-dimensional model, land contours are plotted.

▶ Landsat

Since the Landsat program began in 1972, a number of satellites have been deployed that carry remote sensing equipment. The equipment is designed to detect radiation in different bands of the electromagnetic spectrum. *Landsats 4* and *5* orbit Earth from pole to pole at an altitude of 705 km every 16 days.

Landsat data are particularly useful for thematic mapping. For example, data from the blue-green spectral region are useful for distinguishing between coniferous and deciduous plants, while data from the thermal infrared range supply information about soil moisture.

SECTION 3

Topographic Maps

▶ Inuit Relief Maps

The Inuit of Baffin Island were skilled mapmakers. They made permanent relief maps by carving coastal features into pieces of wood and walrus ivory. The Inuit also sewed small pieces of fur or driftwood to sealskin to represent islands. They measured distance on their maps not by miles but by "sleeps." This distance to a hunting ground, for example, would be measured by how many rest stops would be taken before reaching it.

▶ John Wesley Powell (1834–1902)

American geologist and surveyor John Wesley Powell headed an official expedition to the Grand Canyon in 1871. His purpose was to conduct a topographic survey to map "as broad a belt of country as it was possible" on both sides of the Colorado and Green Rivers. The expedition yielded meticulously detailed topographic maps for an area that had previously been described as the "great unknown." Those maps were instrumental to Powell's appointment in 1881 as director of the U.S. Geological Survey (USGS). As director, Powell aimed to create topographic maps for the entire country using very large scales, ranging from 6.4 km/2.5 cm for desert regions to 1.6 km/2.5 cm for densely populated areas. Powell insisted on including data concerning soils, springs, and other natural resources, which he felt were essential for making land-use decisions. The high-quality topographical maps created during Powell's administration set the standard for published topographic maps in the United States for many years to come.

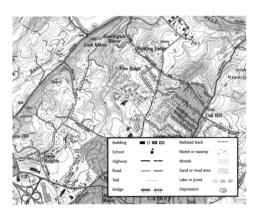

IS THAT A FACT!

▶ By the time Powell retired from the USGS in 1894, about one-fifth of the United States had been mapped according to his standards. Powell told Congress that he expected the remainder of the task (excluding the mapping of Alaska) could be completed in 24 years. However, he vastly underestimated the undertaking; it was not until the 1980s that a full set of topographic maps was finally completed.

For background information about teaching strategies and issues, refer to the *Professional Reference for Teachers*.

CHAPTER 1

Maps as Models of the Earth

 Pre-Reading Questions

Students may not know the answers to these questions before reading the chapter, so accept any reasonable response.

Suggested Answers

1. No; maps do not represent the world accurately because all maps are distorted.

2. Answers may vary. Sample answer: Symbols are used to show information on maps.

3. Every map must have a title, legend, scale, date, and north arrow.

CHAPTER 1

 **Pre-Reading Questions**

1. Do all maps picture the world accurately?

2. How is information shown on maps?

3. What information must every map have?

2

Maps as Models of the Earth

A PICTURE OF THE WORLD

During ancient times, maps of the world were often based on imagination, guesswork, and travelers' tales. Areas of the world that had not yet been visited and explored were sometimes filled in with scenes of mythical places and monsters. Today computer technology and satellite images are used to make maps that are very accurate. In this chapter, you will learn about different kinds of maps and what goes into making a map.

---

internet **connect**

 HRW On-line Resources

go.hrw.com

For worksheets and other teaching aids, visit the HRW Web site and type in the keyword: **HSTMAP**

 SCLINKS NSTA

www.scilinks.com

Use the *sci*LINKS numbers at the end of each chapter for additional resources on the **NSTA** Web site.

 Smithsonian Institution®

www.si.edu/hrw

Visit the Smithsonian Institution Web site for related on-line resources.

 CNN fyi.com

www.cnnfyi.com

Visit the CNN Web site for current events coverage and classroom resources.

START-UP
Activity

FOLLOW THE YELLOW BRICK ROAD

In this activity, you not only will learn how to read a map but also make a map that someone else can read.

Procedure

1. With a **computer drawing program** or **colored pencils** and **paper,** draw a map showing how to get from your classroom to another place in your school, such as the gym. Make sure you include enough information for someone unfamiliar with your school to find his or her way.

2. After you finish drawing your map, switch maps with a partner. Examine your classmate's map, and try to figure out where the map is leading you.

Analysis

3. Is your map an accurate picture of your school? Explain your answer.

4. How do you think your map could be made better? What are some limitations of your map?

5. Compare your map with your partner's map. How are your maps alike? How are they different?

3

START-UP
Activity

FOLLOW THE YELLOW BRICK ROAD

MATERIALS
• computer drawing program
• colored pencils
• paper

Teacher's Notes

Before students start their maps, have them brainstorm to make a list of school landmarks and suggest that they use the location of these landmarks as reference points in their maps.

Answers to START-UP Activity

3. Answers will vary. Accept all reasonable responses.

4. Answers will vary. Accept all reasonable responses.

5. Answers will vary.

Focus

You Are Here

This section opens with a discussion of the history of mapmaking. Students learn how to find directions on a globe by using reference points such as the North and South Poles and the equator. Students learn how a compass is used to find directions and how true north differs from magnetic north. The section closes with a discussion of lines of latitude and longitude and how they can be used to locate points on Earth's surface.

 **Bellringer**

Ask students to draw a map from their homes to one of their favorite places. Have them clearly label all landmarks and include information that might be useful to someone using the map.

1) Motivate

GROUP ACTIVITY

Making Compasses This activity will work best when it is performed outside. Supply each group with a small bowl of water, a steel sewing needle, a magnet, and a 1 × 3 cm piece of tissue paper. Have students carefully rub the needle against the magnet in the same direction 40 times. Then have them float the paper on the surface of the water and place the needle on the paper. After a minute, allow students to observe other groups' bowls; all of the needles should be oriented in a north-south direction. Explain that they have all just created a simple compass, a device that changed the course of human history.

Terms to Learn

map	magnetic declination
true north	longitude
equator	prime meridian
latitude	

What You'll Do

◆ Explain how a magnetic compass can be used to find directions on the Earth.
◆ Distinguish between true north and magnetic north.
◆ Distinguish between lines of latitude and lines of longitude on a globe or map.
◆ Explain how latitude and longitude can be used to locate places on Earth.

You Are Here

When you walk across the Earth's surface, the Earth does not appear to be curved. It looks flat. In the past, beliefs about the Earth's shape changed. Maps reflected the time's knowledge and views of the world as well as the current technology. A **map** is a model or representation of the Earth's surface. If you look at Ptolemy's world map from the second century, as shown in **Figure 1,** you probably will not recognize what you are looking at. Today satellites in space provide us with true images of what the Earth looks like. In this section you will learn how early scientists knew the Earth was round long before pictures from space were taken. You will also learn how to determine location and direction on the Earth's surface.

What Does the Earth Really Look Like?

The Greeks thought of the Earth as a sphere almost 2,000 years before Christopher Columbus made his voyage in 1492. The observation that a ship sinks below the horizon as it sails into the distance supported the idea of a round Earth. If the Earth were flat the ship would appear smaller as it moved away.

Figure 1 *This map shows what people thought the world looked like 800 years ago.*

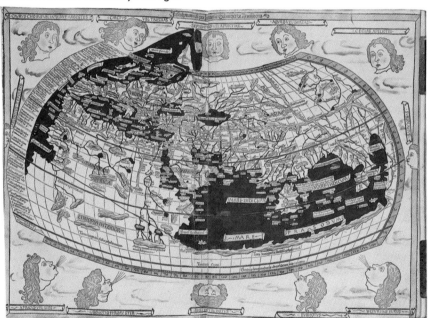

Multicultural CONNECTION

The Chinese invented the magnetic compass in the third century B.C. The magnetic properties of lodestone, known today as magnetite, were well known to many ancient cultures. Lodestone was thought to have magical properties because it could attract metal. Early compasses consisted of a piece of lodestone on a card that was balanced on a pivot. The scale of the compass was marked with compass points at 15° increments. By the tenth century, the compass was standard equipment on Chinese sailing ships.

Eratosthenes (ER uh TAHS thuh NEEZ), a Greek mathematician, wanted to know how big the Earth was. In about 240 B.C., he calculated the Earth's circumference using geometry and observations of the sun. We now know his estimation was off by only 6,250 km, an error of 15 percent. That's not bad for someone who lived more than 2,000 years ago, in a time when computer and satellite technology did not exist!

Finding Direction on Earth

How would you give a friend from school directions to your home? You might mention a landmark, such as a grocery store or a restaurant, as a reference point. A *reference point* is a fixed place on the Earth's surface from which direction and location can be described.

Because the Earth is round, it has no top, bottom, or sides for people to use as reference points for determining locations on its surface. The Earth does, however, turn on its axis. The Earth's axis is an imaginary line that runs through the Earth. At either end of the axis is a geographic pole. The North and South Poles, as shown in **Figure 2,** are used as reference points when describing direction and location on Earth.

Cardinal Directions   North, south, east, and west are called *cardinal directions.* **Figure 3** shows these basic cardinal directions and various combinations of these directions. Using these directions is much more precise than using directions such as turn left, go straight, and turn right. Unfortunately for most of us, using cardinal directions requires the use of a compass.

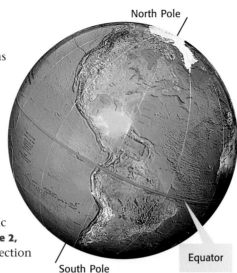

North Pole

Equator

South Pole

Figure 2   *Like the poles, the equator can be used as a reference.*

Figure 3   *A compass rose helps you orient yourself on a map.*

Although globes are round, Earth is not a perfect sphere. Because Earth is a rotating body, it bulges at the equator and is slightly flattened at the poles. Earth's circumference measured pole to pole is 40,008 km, while its circumference around the equator is 40,075 km.

GROUP ACTIVITY

Have students write a description of the route they take as they travel between home and school. Then have them rewrite the description using cardinal and intercardinal directions. Have pairs of students trade route descriptions and use a community map to check the accuracy of each other's maps.
Sheltered English

MISCONCEPTION //ALERT\\\

Students may be under the impression that Christopher Columbus discovered that Earth was round only after he safely made his voyage in 1492 without sailing off the edge of the world. In fact, Columbus, like most other educated people of his time, was well aware that Earth was not flat before he set out.

CONNECT TO
LIFE SCIENCE

Magnetotactic bacteria use magnetic particles in their cytoplasm to align themselves with the Earth's magnetic field. North of the equator, the bacteria are north-seeking travelers. South of the equator, they are south-seeking travelers. At the equator, where the magnetic fields are at their weakest, the populations are mixed between north- and south-seeking bacteria.

   Teaching Transparency 104 "Finding Direction on Earth"

   Directed Reading Worksheet Section 1

READING 📖 STRATEGY

Prediction Guide Before students read the following two pages, have them answer these true/false statements:

- Earth has four poles. (true)
- A compass is the only thing you need to find a location. (false)
- Imaginary lines drawn around Earth can be used to pinpoint locations. (true)

Sheltered English

Answer to Self-Check

The Earth rotates around the geographic poles.

MISCONCEPTION ///ALERT\\\

Explain that magnetic attraction occurs between the opposite poles of magnets. The north-seeking pole of a compass needle actually points to the south pole of Earth's magnetic field. What we call the Earth's magnetic north pole is actually the south pole of Earth's magnetic field. This can be easily demonstrated by placing the south pole of a bar magnet next to a compass. The compass needle will point toward the south pole of the bar magnet in the same way it tends to point toward the south pole of Earth's magnetic field, what we call magnetic north.

 **PG 90**
Orient Yourself!

Finding Directions with a Compass

1. This lab should be done outside. Hold a **compass** flat in your hand until the needle stops moving. Rotate the dial of the compass until the letter *N* on the case lines up with the painted or colored end of the needle.

2. While holding the compass steady, identify objects that line up with each of the cardinal points. List them in your ScienceLog.

3. See if you can locate objects at the various combinations of cardinal directions, such as SW and NE. Record your observations in your ScienceLog.

 TRY at HOME

It's better than a scavenger hunt! Interested? Turn to page 90 of your LabBook.

Using a Compass One way to determine north is by using a magnetic compass. The compass uses the natural magnetism of the Earth to indicate direction. A compass needle points to the magnetic north pole. The Earth has two different sets of poles—the geographic poles and the magnetic poles. As you can see in **Figure 4,** the magnetic poles have a slightly different location than the geographic poles.

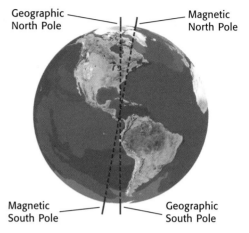

Geographic North Pole — Magnetic North Pole

Magnetic South Pole — Geographic South Pole

Figure 4 *Unlike the geographic poles, which are always in the same place, the magnetic poles have changed location throughout the history of the Earth.*

✔ Self-Check

Does the Earth rotate around the geographic poles or the magnetic poles? *(See page 120 to check your answer.)*

True North and Magnetic Declination Because the geographic North Pole never changes, it is called **true north.** The difference between the location of true north and the magnetic north pole requires that one more step be added to using a compass. When using a compass to map or explore the Earth's surface, you need to make a correction for the difference between geographic north and magnetic north. This angle of correction is called **magnetic declination.** Magnetic declination is measured in degrees east or west of true north.

Magnetic declination has been determined for different points on the Earth's surface. Once you know the declination for your area, you can use a compass to determine true north.

SCIENCE HUMOR

Before the seventeenth century, many sailors refused to transport onions and garlic because they believed they would destroy a compass's magnetic properties. In 1600, English physician and scientist William Gilbert set about testing the belief. He ate a quantity of garlic and then belched on a compass needle, which he had also rubbed with garlic juice. The compass's magnetic properties remained intact, and Gilbert proved, at least to himself, that the notion was unfounded.

This adjustment is like the adjustment you would make to the handlebars of a bike with a bent front wheel. You know how much you have to turn the handlebars to make the bike go straight.

As **Figure 5** shows, a compass needle at Pittsburgh, Pennsylvania, points 5° west of true north. Can you determine the magnetic declination of San Diego?

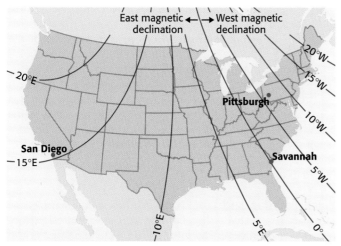

Figure 5 *The red lines on the map connect points with the same magnetic declination.*

Finding Locations on the Earth

The houses and buildings in your neighborhood all have addresses that identify their location. But how would you find the location of something like a city or an island? These places can be given an "address" using *latitude* and *longitude*. Latitude and longitude are intersecting lines on a globe or map that allow you to find exact locations. They are used in combination to create global addresses.

Latitude Imaginary lines drawn around the Earth parallel to the equator are called lines of latitude, or *parallels*. The **equator** is a circle halfway between the poles that divides the Earth into the Northern and Southern Hemispheres. It represents 0° latitude. **Latitude** is the distance north or south, measured in degrees, from the equator, as shown in **Figure 6**. The North Pole is 90° north latitude, and the South Pole is 90° south latitude.

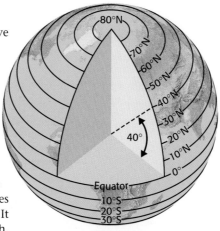

Figure 6 *The degree measure of latitude is the angle created by the equator, the center of the Earth, and the location on the Earth's surface.*

(7)

Astronomy CONNECTION

There are ways you can find north without using a compass. In the morning, the sun rises in the east. In the evening, it sets in the west. If you point your right hand toward where you saw the sun rise in the morning and your left hand to where it sets at night, you will be facing north.

GROUP ACTIVITY

Finding True North For thousands of years, people have used the sun to determine true north. This activity is most accurate at midday, when the sun is at its southernmost point.

- At 11:30 A.M., have students insert a ruler in the ground and a pencil at the tip of the ruler's shadow.
- After 1 hour, have students insert another pencil where the tip of the ruler's shadow is now.
- Have students place a piece of string between the two pencils. The string will be an east-west line. Viewed with your back to the sun, the first pencil indicates west and the second indicates east.

Students can then use a protractor to position a north-south string perpendicular to the east-west string.

Multicultural CONNECTION

Matthew Henson (1866–1955), an African-American member of Robert Peary's North Pole exploration team, was one of the first people believed to have reached the geographic North Pole in 1909. Henson's knowledge of Arctic regions also helped him gain a position as an exhibit preparator at the American Museum of Natural History, in New York City. In 1944, Congress recognized Henson's contributions to polar exploration. Interested students may want to read Henson's autobiography, *A Negro Explorer at the North Pole*.

WEIRD SCIENCE

As the Earth rotates, both the geographic North Pole and the magnetic north pole move constantly. The geographic North Pole moves about 6 m on a 435-day cycle. This movement results from a wobble in the Earth's rotation. The magnetic north pole wanders because of changes in Earth's rotating iron core. The magnetic pole is currently moving northwest at an average rate of 10 km per year.

 Teaching Transparency 105 "Lines of Latitude"

GOING FURTHER

Ask students to find out why the meridian passing through Greenwich, England, serves as the prime meridian. They should find that Greenwich was chosen as 0° by an international committee in 1884 in part because it was the site of Britain's Royal Greenwich Observatory, which had been important in developing time-keeping methods necessary for ship navigation. In addition, most of the world's shipping lines were already using Greenwich as the longitudinal baseline.

Before 1884, when Greenwich was established by international agreement as the prime meridian, there were no fewer than 13 "prime meridians" in use. Countries simply selected a meridian in their own country as the prime meridian; for example, Italy's prime meridian passed through Rome, and France's prime meridian passed through Paris. Discuss with students the benefits of the international standardization of the prime meridian.

Homework

Research Explain to students that longitude is directly related to time. Earth rotates 360° every 24 hours, turning 15° every hour. Earth can be divided into 24 meridians of 15° each. Have students find out how time in different parts of the world can be determined using lines of longitude and the local time at Greenwich.

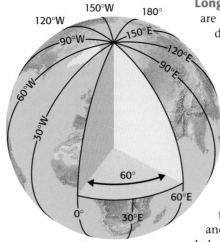

Figure 7 *The degree measure of longitude is the angle created by the prime meridian, the center of the Earth, and the location on the Earth's surface.*

Longitude Imaginary lines that pass through the poles are called lines of longitude, or *meridians*. **Longitude** is the distance east and west, measured in degrees, from the prime meridian, as shown in **Figure 7.** By international agreement, one meridian was selected to be 0°. The **prime meridian,** which passes through Greenwich, England, is the line that represents 0° longitude. Unlike lines of latitude, lines of longitude are not parallel. They touch at the poles and are farthest apart at the equator.

The prime meridian does not completely circle the globe like the equator does. It runs from the North Pole through Greenwich, England, to the South Pole. The 180° meridian lies on the opposite side of the Earth from the prime meridian. Together, the prime meridian and the 180° meridian divide the Earth into two equal halves—the Eastern and Western Hemispheres. East lines of longitude are found east of the prime meridian, between 0° and 180°. West lines of longitude are found west of the prime meridian, between 0° and 180°.

Using Latitude and Longitude Points on the Earth's surface can be located using latitude and longitude. Lines of latitude and lines of longitude intersect, forming a grid system on globes and maps. This grid system can be used to find locations north or south of the equator and east or west of the prime meridian.

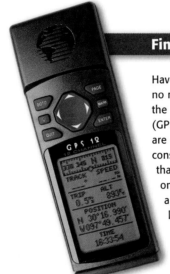

Finding Your Way

Have you ever been lost? There's no need to worry anymore. With the Global Positioning System (GPS), you can find where you are on the Earth's surface. GPS consists of 25 orbiting satellites that send radio signals to receivers on Earth in order to calculate a given location's latitude, longitude, and elevation.

GPS was invented in the 1970s by the United States Department of Defense for military purposes. During the last 20 years, this technology has made its way into many people's daily lives. Today GPS is used in a variety of ways. Airplane and boat pilots use it for navigation, and industry uses include mining and resource mapping as well as environmental planning. Even some cars are equipped with a GPS unit that can display the vehicle's specific location on a computer screen on the dashboard.

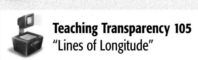

Teaching Transparency 105
"Lines of Longitude"

internet**connect**

SCILINKS
NSTA

TOPIC: Finding Locations on the Earth
GO TO: www.scilinks.org
*sci*LINKS NUMBER: HSTE030

MISCONCEPTION ALERT

Students may believe that a compass is all you need to keep from getting lost outdoors. Point out that compasses are useful only when combined with the ability to read maps and to observe land features. Compasses can be used for orienting oneself and for taking bearings on landmarks for map triangulation.

Figure 8 shows you how latitude and longitude can be used to find the location of your state capital. First locate the star symbol representing your state capital on the appropriate map. Find the lines of latitude and longitude closest to your state capital. From here you can estimate your capital's approximate latitude and longitude.

Activity

Use an atlas or globe to find the latitude and longitude of the following cities:
New York, New York
Sao Paulo, Brazil
Sydney, Australia
Madrid, Spain
Cairo, Egypt

TRY at HOME

Answers to Activity
New York, New York: 40°N, 74°W
Sao Paulo, Brazil: 23°S, 46°W
Sydney, Australia: 33°S, 151°E
Madrid, Spain: 40°N, 3°W
Cairo, Egypt: 30°N, 31°E

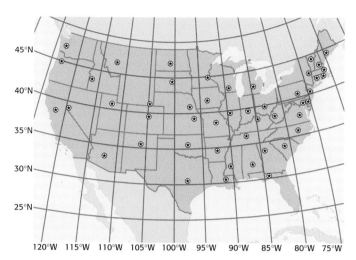

Figure 8 *The grid pattern formed by lines of latitude and longitude allows you to pinpoint any location on the Earth's surface.*

SECTION REVIEW

1. Explain the difference between true north and magnetic north.

2. When using a compass to map an area, why is it important to know an area's magnetic declination?

3. In what three ways is the equator different from the prime meridian?

4. How do lines of latitude and longitude help you find locations on the Earth's surface?

5. **Applying Concepts** While digging through an old trunk, you find a treasure map. The map shows that the treasure is buried at 97° north and 188° east. Explain why this is impossible.

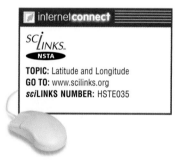

internetconnect

SCiLINKS
NSTA

TOPIC: Latitude and Longitude
GO TO: www.scilinks.org
*sci*LINKS NUMBER: HSTE035

4 Close

Quiz

1. Name three references that can be used to describe direction and location on Earth.
(possible answers: North Pole, South Pole, the equator, lines of latitude and longitude)

2. What are lines of latitude and lines of longitude?
(Lines of latitude are imaginary lines around Earth parallel to the equator, and they are used to measure a location's distance north or south of the equator. Lines of longitude are imaginary lines that run between the Earth's poles, and they are used to measure a location's distance east or west of the prime meridian.)

ALTERNATIVE ASSESSMENT

Have students use a world map to plan an around-the-world trip in which they give their various destinations only in degrees of latitude and longitude. Have students trade their itinerary with a partner, and have each "decode" the other's trip.

Reinforcement Worksheet
"Where on Earth?"

internetconnect

SCiLINKS
NSTA

TOPIC: Latitude and Longitude
GO TO: www.scilinks.org
*sci*LINKS NUMBER: HSTE035

9

▼ Answers to Section Review

1. Because the geographic North Pole never changes, it is called true north. Magnetic north refers to the magnetic north pole, which has changed throughout history.

2. Because the compass points to magnetic north, it is important to know the magnetic declination at your location. This will help you make corrections to adjust for the difference between true north and magnetic north.

3. Answers will vary.

4. Lines of latitude and lines of longitude form a grid system that can be used to find locations on the Earth's surface.

5. This is impossible because the greatest measure of latitude is 90° and the greatest measure of longitude is 180°.

Focus

Mapping the Earth's Surface

In this section, students compare the uses of maps and globes and explore the features of three common map projections. In addition, they learn the parts of a map and discover some of the technological advances that have influenced recent trends in cartography.

Bellringer

Display a world map, a map of your state, and a map of your community. Have students make a chart in which they list the similarities and differences between each map. Then have students suggest three uses for each map.

1) Motivate

DISCUSSION

Have students examine a globe and a Mercator projection of a world map. Point out to students that both are representations of Earth. Then challenge students to find examples of ways in which the two representations differ. If necessary, point out the relative difference in the size and shape of Greenland. Ask students how they might account for the discrepancies. Record students' ideas on the chalkboard, and tell them that in this section they will learn about some of the difficulties involved in making flat representations of Earth's curved surface.

Directed Reading Worksheet Section 2

Terms to Learn

Mercator projection
conic projection
azimuthal projection
aerial photograph
remote sensing

What You'll Do

◆ Compare a map with a globe.
◆ Describe the three types of map projections.
◆ Describe recent technological advances that have helped the science of mapmaking progress.
◆ List the parts of a map.

Mapping the Earth's Surface

Models are often used to represent real objects. For example, architects use models of buildings to give their clients an idea of what a building will look like before it is completed. Likewise, Earth scientists often make models of the Earth. These models are globes and maps.

Because a globe is a sphere, a globe is probably the most accurate model of the Earth. Also, a globe accurately represents the sizes and shapes of the continents and oceans in relation to one another. But a globe is not always the best model to use when studying the Earth's surface. For example, a globe is too small to show a lot of detail, such as roads and rivers. It is much easier to show details on maps. Maps can show the entire Earth or parts of it. But how do you represent the Earth's curved surface on a flat surface? Read on to find out.

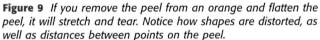

A Flat Sphere?

A map is a flat representation of the Earth's curved surface. However, when you transfer information from a curved surface to a flat surface, you lose some accuracy. Changes called distortions occur in the shapes and sizes of landmasses and oceans. These distortions make some landmasses appear larger than they really are. Direction and distance can also be distorted. Consider the example of the orange peel shown in **Figure 9.**

Figure 9 *If you remove the peel from an orange and flatten the peel, it will stretch and tear. Notice how shapes are distorted, as well as distances between points on the peel.*

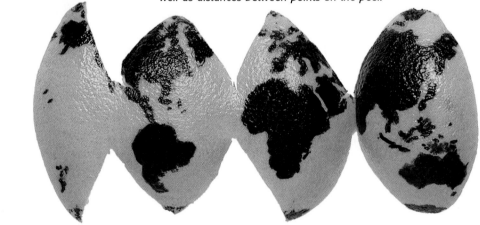

SCIENCE HUMOR

Q: What do you get when you cross a cowboy with a mapmaker?

A: a cowtographer

Map Projections Mapmakers use map projections to transfer the image of Earth's curved surface onto a flat surface. No map projection of the Earth can represent the surface of a sphere exactly. All flat maps have some amount of distortion. A map showing a smaller area, such as a city, has much less distortion than a map showing a larger area, such as the entire world.

To understand how map projections are made, imagine the Earth as a transparent globe with a light inside. If you hold a piece of paper up against the globe, shadows appear on the paper that show markings on the globe, such as continents, lines of latitude, and lines of longitude. The way the paper is held against the globe determines the kind of projection that is made. The most common projections are based on three geometric shapes—cylinders, cones, and planes.

Mercator Projection A **Mercator projection** is a map projection that results when the contents of the globe are transferred onto a cylinder of paper, as shown in **Figure 10.**

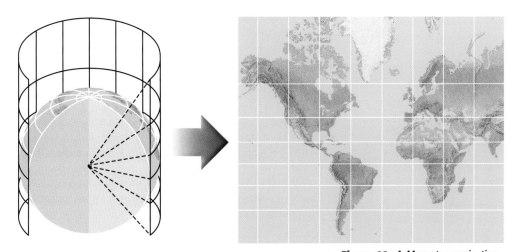

Figure 10 *A Mercator projection is accurate near the equator but distorts distances and sizes of areas near the poles.*

The Mercator projection shows the Earth's latitude and longitude as straight, parallel lines. Lines of longitude are plotted with an equal amount of space between each line. Lines of latitude are spaced farther apart north and south of the equator. Making the lines parallel widens and lengthens the size of areas near the poles. For example, on the Mercator projection in the map shown above, Greenland appears almost as large as Africa. Actually, Africa is 15 times larger than Greenland.

2 Teach

READING STRATEGY

Activity As students read about map projections, have them write down the name of each type of projection and a brief description of what the projection represents. Have students list the advantages or disadvantages of each type of projection. **Sheltered English**

MISCONCEPTION ALERT

The distortions of landmasses are not the only inaccuracies that occur on maps. When making maps for popular use, such as road maps, mapmakers routinely generalize them for both practical and aesthetic reasons. For example, when the size of a map's scale is reduced, two features (such as two lakes or two towns) might appear to be adjacent to each other. In this case, the mapmaker might move them slightly apart. Mapmakers sometimes also add details that may not really exist; for instance, meander loops might be added to a river or stream to make it look more realistic. Topographic maps, however, are made from aerial photographs and are extremely accurate.

 Teaching Transparency 106 "Mercator Projection"

11

Mnemonics Have students think of some rhymes to help them remember key points about the projections discussed in the text. You might suggest the following to help students get started:

"If you're traveling to the equator, you'll do well with Mercator"; "for east to west, conic is best"; "for a stroll at a pole, an azimuthal will help you stay in control." **Sheltered English**

CROSS-DISCIPLINARY FOCUS

History One of the earliest known maps was made by the Babylonians. It shows Babylon at the center, with Syria and other territories represented as a circular area surrounded by the Persian Gulf. This type of map is called a "wheel map." Many cultures in Arabian and European countries used wheel maps. Wheel maps reinforced the idea that a civilization was at the center of the universe. Ask students to draw a wheel map of the area where they live.

Teaching Transparency 106
"Conic Projection"
"Azimuthal Projection"

Conic Projection A **conic projection** is a map projection that is made by transferring the contents of the globe onto a cone, as shown in **Figure 11**. This cone is then unrolled to form a flat plane.

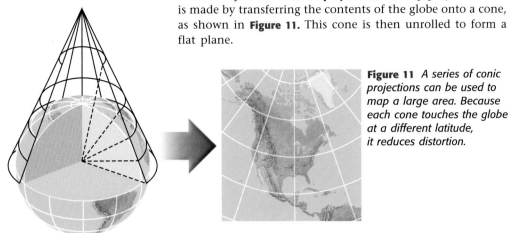

Figure 11 *A series of conic projections can be used to map a large area. Because each cone touches the globe at a different latitude, it reduces distortion.*

The cone touches the globe at each line of longitude but only one line of latitude. There is no distortion along the line of latitude where the globe comes in contact with the cone. Areas near this line of latitude are distorted the least amount. Because the cone touches many lines of longitude and only one line of latitude, conic projections are best for mapping landmasses that have more area east to west, such as the United States, than north to south, such as South America.

Azimuthal Projection An **azimuthal** (AZ i MYOOTH uhl) **projection** is a map projection that is made by transferring the contents of the globe onto a plane, as shown in **Figure 12**.

On an azimuthal projection, the plane touches the globe at only one point. Little distortion occurs at the point of contact, which is usually one of the poles. However, distortion of direction, distance, and shape increases as the distance from the point of contact increases.

Figure 12 *On this azimuthal projection, distortion increases as you move further away from the North Pole.*

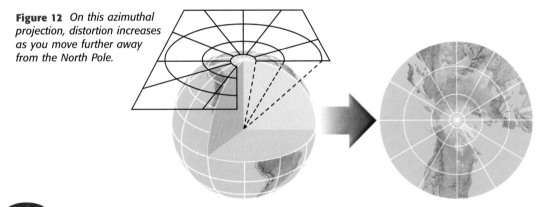

Homework

Map Projections Have students make a chart listing the strengths and weaknesses of each of the three projections studied in this section. Then have them also research another projection, such as the Robinson projection. Have them add the strengths and weaknesses of that projection to their charts. At the bottom of the chart, have them explain why none of the projections is entirely free of distortions and inaccuracies.

Modern Mapmaking

The science of mapmaking has changed more since the beginning of the 1900s than during any other time in history. This has been due to many technological advances in the twentieth century, such as the airplane, photography, computers, and space exploration.

Airplanes and Cameras   The development of the airplane and advancements in photography have had the biggest effect on modern mapmaking. Airplanes give people a bird's-eye view of the Earth's surface. Photographs from the air are called **aerial photographs.** These photographs are important in helping mapmakers make accurate maps.

Remote Sensing   The combined use of airplanes and photography led to the science of remote sensing. **Remote sensing** is gathering information about something without actually being there. Remote sensing can be as basic as cameras in planes or as sophisticated as satellites with sensors that can sense and record what our eyes cannot see. Remotely sensed images allow a mapmaker to map the surface of the Earth more accurately.

Our eyes can detect only a small part of the sun's energy. The part we see is called visible light. Remote sensors on satellites can detect energy that we cannot see. Satellites do not take photographs with film like cameras do. A satellite collects information about energy coming from the Earth's surface and sends it back to receiving stations on Earth. A computer is then used to process the information to create an image we can see, like the one shown in **Figure 13.**

Figure 13 *Satellites can detect objects the size of a baseball stadium. The satellite that took this picture was 220 km above the Earth's surface!*

13

Science Bloopers

During the complex process of compiling vast amounts of data from a number of different maps and sources, mistakes are sometimes made. Cartographers have wiped entire cities off maps accidentally! Canada's capital, Ottawa, for example, was once omitted from a Canadian tourist-office map. An official explanation that there was no direct air service between New York City and Ottawa failed to satisfy one Ottawa tourist bureau executive, who remarked irately, "Ottawa should be shown in any case, even if the only point of entry was by two-man kayak."

Learners Having Difficulty

Show students two maps with different representative fraction scales, such as a map of North America and a map of your community. Point out that the larger the denominator in a representative fraction scale, the smaller the scale of the map. You might clarify this by pointing out that $\frac{1}{4}$ of a pie is smaller than $\frac{1}{2}$ of a pie. Maps that show a large area, such as a continent, use a smaller scale and show less detail. Maps that show a smaller area, such as a town, can have a larger scale and show more detail. Discuss with students cases in which large-scale maps are the most useful and cases in which small-scale maps are most useful.

Sheltered English

MATH and MORE

Have students suppose that they want to use a map with a scale of 1:24,000 to estimate the length of a hike. On the map, the route measures 20 cm. Have students calculate the length of the hike in kilometers.

(20 cm × 24,000 = 480,000 cm;
480,000 cm ÷ 100 cm/m = 4,800 m;
4,800 ÷ 1,000 m/km = 4.8 km)

Math Skills Worksheet
"Using Proportions and Cross-Multiplication"

Critical Thinking Worksheet
"Shaping the World"

Information Shown on Maps

As you have already learned, there are many different ways of making maps. It is also true that there are many types of maps. You might already be familiar with some, such as road maps or political maps of the United States. But regardless of its type, each map should contain the information shown in **Figure 14.**

Figure 14 Road Map of Connecticut

The **title** tells you what area is being shown on the map or gives you information about the subject of the map.

A **map's scale** shows the relationship between the distance on the Earth's surface and the distance on the map.

A **graphic scale** is like a ruler. The distance on the Earth's surface is represented by a bar graph that shows units of distance.

A **verbal scale** is a phrase that describes the measure of distance on the map relative to the distance on the Earth's surface.

A **representative fraction** is a fraction or ratio that shows the relationship between the distance on the map and the distance on the Earth's surface. It is unitless, meaning it stays the same no matter what units of measurement you are using.

14

Multicultural CONNECTION

Maps made by Native Americans were similar to maps made by Europeans in that they included not only geographic information but also elements of their history, traditions, and mythology. Native American mapmakers also used standardized symbols to indicate roads, villages, rivers, and other physical features. The Aztec of central Mexico, for example, used rows of footprints to represent roads and swirling lines to indicate water. Aztec maps and mapmaking techniques were adapted by the Spanish, who explored and conquered the region.

Reading a Map

Imagine that you are a trip planner for an automobile club. A couple of people come in who want to travel from Torrington, Connecticut, to Bristol, Connecticut. Using the map in Figure 14, describe the shortest travel route you would suggest they take between the two cities. List the roads they would take, the direction they would travel, and the towns they would pass through. Use the map scale to determine approximately how many miles there are between Torrington and Bristol.

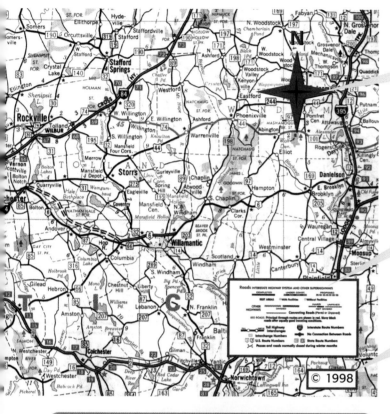

A **compass rose** shows you how the map is positioned in relation to true north.

A **legend** is a list of the symbols used in the map and their explanations.

The **date** gives the time at which the information on the map was accurate.

© 1998

SECTION REVIEW

1. A globe is a fairly accurate model of the Earth, yet it has some weaknesses. What is one weakness?

2. What is distortion on a map, and why does it occur?

3. What is remote sensing? How has it changed mapmaking?

4. **Summarizing Data** List five items found on maps. Explain how each item is important to reading a map.

internetconnect

SCI**LINKS**
NSTA

TOPIC: Mapmaking
GO TO: www.scilinks.org
*sci*LINKS NUMBER: HSTE040

15

Focus

Topographic Maps

In this section, students investigate how contour lines are used to show elevation and landforms on a topographic map. In addition, they learn how to read and interpret the features of a topographic map.

 **Bellringer**

Have students examine the topographic map shown in **Figure 15**. Have them imagine that they are standing on the top of Campbell Hill. Students should describe in their ScienceLog what they see in each direction. Tell students that they will learn to read topographic maps, such as the ones in this section.

1 Motivate

ACTIVITY

Investigate Your Area If possible, obtain topographic maps of your area from the USGS. Display the maps for students to study. As a class, locate different landforms, such as lakes, mountains, and valleys. Discuss with students how contour intervals indicate changes in elevation. If possible, take a class field trip to an area shown on a topographic map. Students will enjoy comparing the map with the topography they observe.
Sheltered English

 Directed Reading Worksheet Section 3

Terms to Learn

topographic map
elevation
contour lines
contour interval
relief
index contour

What You'll Do

◆ Describe how contour lines show elevation and landforms on a map.
◆ List the rules of contour lines.
◆ Interpret a topographic map.

Topographic Maps

Imagine that you are on an outdoor adventure trip. The trip's purpose is to improve your survival skills by having you travel across undeveloped territory with only a compass and a map. What kind of map will you be using? Well, it's not a road map—you won't be seeing a lot of roads where you are going. You will need a topographic map. A **topographic map** is a map that shows surface features, or topography, of the Earth. Topographic maps show both natural features, such as rivers, lakes and mountains, and features made by humans, such as cities, roads, and bridges. Topographic maps also show elevation. **Elevation** is the height of an object above sea level. The elevation at sea level is 0. In this section you will learn how to interpret a topographic map.

Elements of Elevation

The United States Geological Survey (USGS), a federal government agency, has made topographic maps for all of the United States. Each of these maps is a detailed description of a small area of the Earth's surface. Because the topographic maps produced by the USGS use feet as their unit of measure rather than meters, we will follow their example.

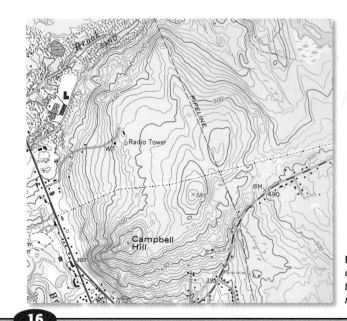

Contour Lines On a topographic map, contour lines are used to show elevation. **Contour lines** are lines that connect points of equal elevation. For example, one contour line would connect points on a map that have an elevation of 100 ft. Another line would connect points on a map that have an elevation of 200 ft. **Figure 15** illustrates how contour lines appear on a map.

Figure 15 *Because contour lines connect points of equal elevation, the shape of the contour lines reflects the shape of the land.*

16

IS THAT A FACT!

The Ordnance Survey of Great Britain produces topographic maps with very large scales, ranging from 1:10,000 to 1:1,250. Such large scales permit a level of detail that extends to showing the location of public telephones, windmills, and large boulders!

Contour Interval The difference in elevation between one contour line and the next is called the **contour interval.** For example, a map with a contour interval of 20 ft would have contour lines every 20 ft of elevation change, such as 0 ft, 20 ft, 40 ft, 60 ft, and so on. A mapmaker chooses a contour interval based on the area's relief. **Relief** is the difference in elevation between the highest and lowest points of the area being mapped. Because the relief of a mountainous area is high, it might be shown on a map using a large contour interval, such as 100 ft. However, a flat area has low relief and might be shown on a map using a small contour interval, such as 10 ft.

The spacing of contour lines also indicates slope, as shown in **Figure 16.** Contour lines that are close together, with little space between them, usually show a steep slope. Contour lines that are spaced far apart generally represent a gentle slope.

Index Contour On many topographic maps, the mapmaker uses an index contour to make reading the map a little easier. An **index contour** is a darker, heavier contour line that is usually every fifth line and that is labeled by elevation. Find an index contour on both of the topographic maps.

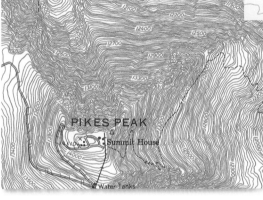

Figure 16 *The portion of the topographic map on the left shows Pikes Peak, in Colorado. The map above shows a valley in Big Bend Ranch State Park, in Texas.*

✓ Self-Check

If elevation is not labeled on a map, how can you determine if the mapped area is steep or not?
(See page 120 to check your answer.)

17

MATH BREAK

Counting Contours

Calculate the contour interval for the map shown in Figure 15 on the previous page. (Hint: Find the difference between two bold lines found next to each other. Subtract the lower marked elevation from the higher marked elevation. Divide by 5.)

② Teach

Answers to MATHBREAK

Sample answer:
500 ft − 450 ft = 50 ft
50 ft ÷ 5 = 10 ft
contour interval = 10 ft

DISCUSSION

Reproduce on the chalkboard a mountainous portion of one of the contour maps shown in the text. Discuss with students where the steepest slopes are (where the lines are closest together). Then change the contour interval by erasing every other contour line. Point out to students that the contour interval is now twice as large. Discuss with students the advantages and the disadvantages of a map with a larger contour interval. (The map may seem easier to read, but detail is lost.)

RETEACHING

If students are having trouble understanding contour lines, use some modeling clay to make a landform. Ask a volunteer to hold a ruler vertically next to the landform while you use a plastic knife to mark off contour intervals around the landform. When you are finished, have students view the landform from the side so that they can see the uniformity of the contour intervals and from above so they can see how the intervals would appear on a topographic map. Then give pairs of students some modeling clay and invite them to try the same activity themselves. **Sheltered English**

Answers to Self-Check

If the lines are close together, then the mapped area is steep. If the lines are far apart, the mapped area has a gradual slope or is flat.

Homework

Expedition Journal Have students write a fictitious journal from the perspective of a member of an expedition team. Every journal entry should include a description of the topography they encountered. Students should include a map in which they use the appropriate symbols and contour lines to show their route. This activity will take several days to complete. After the student explorers have "returned" from their expedition, they should present their map and read their journal entries to the class.

3) Extend

GUIDED PRACTICE

Display a portion of a topographic map and point to the following features. Have students identify them and describe what they indicate.

- index contour line (identifiable from its color and its label elevation; index contours make reading the map easier)
- steep slope (identifiable by the close spacing of the contour lines)
- gentle slope (identifiable because the contour lines are relatively far apart)
- river, lake, or pond (identifiable by shape and color)

Sheltered English

GROUP ACTIVITY

Have groups put together a contour map reading presentation for another class. Encourage them to make some handouts for the class about topographic maps and their uses and to include visual elements, such as a poster or a diagram, that highlight the features of topographic maps. Groups might finish their presentation with a quiz to assess how well they've explained the material.

MEETING INDIVIDUAL NEEDS

Advanced Learners Have students research orienteering and learn some of the techniques involved. Encourage students to explain or demonstrate for classmates such skills as how to set a map with a compass, how to determine bearings, and how to reconcile the differences between magnetic north, grid north, and true north.

Reading a Topographic Map

Topographic maps, like other maps, use symbols to represent parts of the Earth's surface. The legend from the USGS topographic map in **Figure 17** shows some of the common symbols used to represent certain features in topographic maps.

Different colors are also used to represent different features of the Earth's surface. In general, buildings, roads, bridges, and railroads are black. Contour lines are brown. Major highways are red. Cities and towns are pink. Bodies of water, such as rivers, lakes, and oceans, are shown in blue, and wooded areas are represented by the color green.

Figure 17 *All USGS topographic maps use the same legend to represent natural features and features made by humans.*

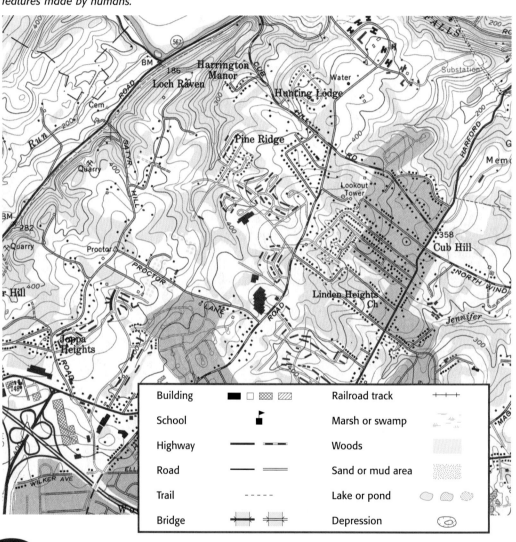

18

CONNECT TO OCEANOGRAPHY

Oceanographers use contour maps to map the topography of the ocean floor. Traditionally, darker colors represent deeper depths, while lighter colors represent areas closer to the surface of the water. If possible, display an oceanographic map, and have students apply what they have learned about topographic maps to create a profile of a section of the ocean floor. Students should find similarities between the topography of the ocean floor and that of continental landmasses.

The Golden Rules of Contour Lines Contour lines are the key to interpreting the size and shape of landforms on a topographic map. When you first look at a topographic map, it might seem confusing. Accurately reading a topographic map requires training and practice. The following rules will help you understand how to read topographic maps:

1. Contour lines never cross. All points along a contour line represent a single elevation.

2. The spacing of contour lines depends on slope characteristics. Closely spaced contour lines represent a steep slope. Widely spaced contour lines represent a gentle slope.

3. Contour lines that cross a valley or stream are V-shaped. The V points toward the area of higher elevation. If a stream or river flows through the valley, the V points upstream.

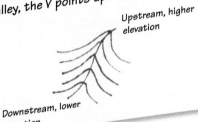

Upstream, higher elevation

Downstream, lower elevation

4. Contour lines form closed circles around the tops of hills, mountains, and depressions. One way to tell hills and depressions apart is that depressions are marked with short, straight lines inside the circle, pointing downslope toward the center of the depression.

Hill

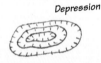

Depression

Environment CONNECTION

State agencies, such as the Texas Parks and Wildlife Department, use topographic maps to plot the distribution and occurrence of endangered plant and animal species. By marking the location of endangered species on a map, these agencies can record and protect these habitats.

SECTION REVIEW

1. How do topographic maps represent the Earth's surface?

2. If a contour map contains streams, can you tell where the higher ground is even if all of the numbers are removed?

3. Why can't contour lines cross?

4. **Inferring Conclusions** Why isn't the highest point on a hill or a mountain represented by a contour line?

internet**connect**

sci**LINKS** NSTA

TOPIC: Topographic Maps
GO TO: www.scilinks.org
sci**LINKS NUMBER:** HSTE045

19

LabBook PG 92
Topographic Tuber

Quiz

1. What is a contour interval on a topographic map? (the difference in elevation between one contour line and the next)

2. What do closely spaced contour lines on a topographic map indicate? (a steep area)

3. How does a topographic map indicate the direction that a stream flows? (Streams flow downhill, or in the direction that elevation decreases. The Vs point toward higher elevations.)

ALTERNATIVE ASSESSMENT

Photocopy a portion of a topographic map that shows a mountain. Distribute copies to students, and ask them to trace a route to the top. Then have them write a description of their "trail" that describes the length of the hike, at which elevations the trail is the steepest, and at which points the slope is gentle. Have them also note other features such as streams, power lines, or road crossings.

Reinforcement Worksheet
"Interpreting a Topographic Map"

Math Skills Worksheet
"Mapping and Surveying"

▼ **Answers to Section Review**

1. Topographic maps use contour lines to show the surface features of Earth.

2. Yes; contour lines that cross a stream are V-shaped. The V points toward the area of higher elevation.

3. Contour lines cannot cross because they represent a certain elevation. If contour lines crossed, the point where the lines crossed would have two different elevations, which is impossible.

4. The highest point on a hill or mountain is a single point, not a group of points with the same elevation.

Skill Builder Lab

USING SCIENTIFIC **METHODS**

Round or Flat?
Teacher's Notes

Round or Flat?

Eratosthenes thought he could measure the circumference of Earth. He came up with the idea while reading that a well in southern Egypt was entirely lit by the sun at noon once each year. He realized that for this to happen, the sun must be directly over the well! But at the same time, in a city just north of this well, a tall monument cast a shadow. Eratosthenes reasoned that the sun could not be directly over both the monument and the well at noon on the same day. In this experiment, you will test his idea and see for yourself how his investigation works.

Time Required

One 45-minute class period

Lab Ratings

EASY ———————————→ HARD

TEACHER PREP   ⚗⚗
STUDENT SET-UP ⚗⚗⚗
CONCEPT LEVEL   ⚗⚗⚗
CLEAN UP   ⚗

MATERIALS

The materials listed on the student page are enough for a group of 3–4 students.

Safety Caution

Remind students to review all safety cautions before beginning this lab activity.

Preparation Notes

Obtain inflated basketballs from your school's physical education instructor. It may be necessary to ask students to bring basketballs from home. Begin the activity by reminding students that *circumference* is the distance around a circle or sphere.

You may also need to review the use of protractors with students before performing this activity.

Ask a Question

1 How could I use Eratosthenes' investigation to measure the size of Earth?

Form a Hypothesis

2 In your ScienceLog, write a few sentences that answer the question above.

Test the Hypothesis

3 Set the basketball on a table. Place a book or notebook on either side of the basketball to hold the ball in place. The ball represents Earth.

4 Use modeling clay to attach a pencil to the "equator" of the ball so that the pencil points away from the ball.

5 Attach the second pencil to the ball 5 cm above the first pencil. This second pencil should also point away from the ball.

6 Using a meterstick, mark on the table with masking tape a position 1 m away from the ball. Label the position "Sun." Place the flashlight here.

7 When your teacher turns out the lights, turn on your flashlight and point it so that the pencil on the equator does not cast a shadow. Ask a partner to hold the flashlight in this position. The second pencil should cast a shadow on the ball.

MATERIALS

- basketball
- 2 books or notebooks
- modeling clay
- 2 unsharpened pencils
- metric ruler
- meterstick
- masking tape
- flashlight or small lamp
- string, 10 cm long
- protractor
- tape measure
- calculator (optional)

20

CLASSROOM TESTED & APPROVED

Barry L. Bishop
San Rafael Junior High
Ferron, Utah

Lab Notes

Explain that Eratosthenes' experiment worked because he set up a ratio. It may be necessary to review ratios before performing this activity. The formula Eratosthenes used is as follows:

$$\frac{\text{Distance around ball}}{\text{Distance between sticks}} = \frac{360° \text{ in the sphere}}{\text{Angle of shadow with stick}}$$

8 Tape one end of the string to the top of the second pencil. Hold the other end of the string against the ball at the far edge of the shadow. Make sure that the string is tight. Be careful not to pull the pencil over.

9 Use a protractor to measure the angle between the string and the pencil. Record this angle in your ScienceLog.

10 Use the following formula to calculate the *experimental circumference* of the ball:

$$circumference = \frac{360° \times 5\ cm}{angle\ between\ pencil\ and\ string}$$

Record this circumference in your ScienceLog.

11 Wrap the tape measure around the ball's "equator" to measure the *actual circumference* of the ball. Record this circumference in your ScienceLog.

Analyze the Results

12 In your ScienceLog, compare the experimental circumference with the actual circumference.

13 What could have caused your experimental circumference to be different from the actual circumference?

14 What are some of the advantages and disadvantages of taking measurements this way?

Draw Conclusions

15 Was this method an effective way for Eratosthenes to measure Earth's circumference? Explain your answer.

Lab Notes
Students may be interested to learn that they can calculate the circumference of the Earth by performing Eratosthenes' experiment in partnership with other schools around the world. The experiment is conducted twice a year during the fall and spring equinoxes. To find out more, have students search for "Eratosthenes' experiment" on the Internet.

Answers
12. Students are likely to find that the experimental and actual circumference are not equal, but the two values should be close.

13. Answers may vary. Factors that affect this measurement include a slant of the pencils and human error in measurement.

14. Because it is impossible to use a tape measure to determine the circumference of the Earth, this procedure offers a good approximation. One disadvantage is that the measurements are not exact.

15. Yes, because it gives a value that is close to the actual value.

 Datasheets for LabBook

 Science Skills Worksheet "Working with Hypotheses"

Chapter Highlights

Chapter Highlights

VOCABULARY DEFINITIONS

SECTION 1

map model or representation of the Earth's surface

true north the geographic North Pole

magnetic declination the angle of correction for the difference between geographic north and magnetic north

equator a circle halfway between the poles that divides the Earth into the Northern and Southern Hemispheres

latitude the distance north or south from the equator; measured in degrees

longitude the distance east or west from the prime meridian; measured in degrees

prime meridian the line of longitude that passes through Greenwich, England; represents 0° longitude

SECTION 2

Mercator projection a map projection that results when the contents of the globe are transferred onto a cylinder

conic projection a map projection that is made by transferring the contents of the globe onto a cone

azimuthal projection a map projection that is made by transferring the contents of the globe onto a plane

aerial photograph a photograph taken from the air

remote sensing gathering information about something without actually being nearby

SECTION 1

Vocabulary

map (p. 4)
true north (p. 6)
magnetic declination (p. 6)
equator (p. 7)
latitude (p. 7)
longitude (p. 8)
prime meridian (p. 8)

Section Notes

- The North and South Poles are used as reference points for describing direction and location on the Earth.

- The cardinal directions—north, south, east, and west—are used for describing direction.

- Magnetic compasses are used to determine direction on the Earth's surface. The north needle on the compass points to the magnetic north pole.

- Because the geographic North Pole never changes location, it is called true north. The magnetic poles are different from the Earth's geographic poles and have changed location throughout the Earth's history.

- The magnetic declination is the adjustment or difference between magnetic north and geographic north.

- Latitude and longitude are intersecting lines that help you find locations on a map or a globe. Lines of latitude run east-west. Lines of longitude run north-south through the poles.

Labs

Orient Yourself! (p. 90)

SECTION 2

Vocabulary

Mercator projection (p. 11)
conic projection (p. 12)
azimuthal projection (p. 12)
aerial photograph (p. 13)
remote sensing (p. 13)

Section Notes

- A globe is the most accurate representation of the Earth's surface.

- Maps have built-in distortion because some information is lost when mapmakers transfer images from a curved surface to a flat surface.

☑ Skills Check

Math Concepts

REPRESENTATIVE FRACTION One type of map scale is a representative fraction. A representative fraction is a fraction or ratio that shows the relationship between the distance on the map and the distance on the Earth's surface. It is unitless, meaning it stays the same no matter what units of measurement you are using. For example, say you are using a map with a representative fraction scale that is 1:12,000. If you are measuring distance on the map in centimeters, 1 cm on the map represents 12,000 cm on the Earth's surface. A measure of 3 cm on the map represents 12,000 × 3 cm = 36,000 cm on the Earth's surface.

Visual Understanding

THE POLES The Earth has two different sets of poles—the geographic poles and the magnetic poles. See Figure 4 on page 6 to review how the geographic poles and the magnetic poles differ.

INFORMATION SHOWN ON MAPS Study Figure 14 on pages 14 and 15 to review the necessary information each map should contain.

Lab and Activity Highlights

Round or Flat? PG 20

Orient Yourself! PG 90

Topographic Tuber PG 92

Datasheets for LabBook (blackline masters for these labs)

SECTION 2

- Mapmakers use map projections to transfer images of the Earth's curved surface to a flat surface.

- The most common map projections are based on three geometric shapes—cylinders, cones, and planes.

- Remote sensing has allowed mapmakers to make more accurate maps.

- All maps should have a title, date, scale, legend, and north arrow.

SECTION 3

Vocabulary

topographic map (p. 16)

elevation (p. 16)

contour lines (p. 16)

contour interval (p. 17)

relief (p. 17)

index contour (p. 17)

Section Notes

- Topographic maps use contour lines to show a mapped area's elevation and the shape and size of landforms.

- The shape of contour lines reflects the shape of the land.

- The contour interval and the spacing of contour lines indicate the slope of the land.

- Like all maps, topographic maps use a set of symbols to represent features of the Earth's surface.

- Contour lines never cross. Contour lines that cross a valley or stream are V-shaped. Contour lines form closed circles around the tops of hills, mountains, and depressions.

Labs

Topographic Tuber (p. 92)

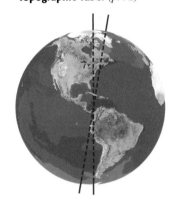

internet connect

GO TO: go.hrw.com

Visit the **HRW** Web site for a variety of learning tools related to this chapter. Just type in the keyword:

KEYWORD: HSTMAP

SCI LINKS
NSTA

GO TO: www.scilinks.org

Visit the **National Science Teachers Association** on-line Web site for Internet resources related to this chapter. Just type in the *sci*LINKS number for more information about the topic:

TOPIC: Finding Locations on the Earth   *sci*LINKS NUMBER: HSTE030
TOPIC: Latitude and Longitude   *sci*LINKS NUMBER: HSTE035
TOPIC: Mapmaking   *sci*LINKS NUMBER: HSTE040
TOPIC: Topographic Maps   *sci*LINKS NUMBER: HSTE045

23

Lab and Activity Highlights

LabBank

 Inquiry Labs, Looking for Buried Treasure

Long-Term Projects & Research Ideas, Globe Trotting

Chapter Review

Explain the difference between the following sets of words:

1. true north/magnetic north
2. latitude/longitude
3. equator/prime meridian
4. Mercator projection/azimuthal projection
5. contour interval/index contour
6. elevation/relief

UNDERSTANDING CONCEPTS

Multiple Choice

7. A point whose latitude is 0° is located on the
   a. North Pole. c. South Pole.
   b. equator. d. prime meridian.

8. The distance in degrees east or west of the prime meridian is
   a. latitude. c. longitude.
   b. declination. d. projection.

9. The needle of a magnetic compass points toward the
   a. meridians.
   b. parallels.
   c. geographic North Pole.
   d. magnetic north pole.

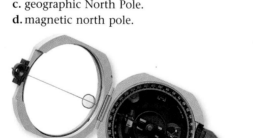

10. The most common map projections are based on three geometric shapes. Which of the following geometric shapes is not one of them?
    a. cylinder c. cone
    b. square d. plane

11. A Mercator projection is distorted near the
    a. equator.
    b. poles.
    c. prime meridian.
    d. date line.

12. What kind of scale does not have written units of measure?
    a. representative fraction
    b. verbal
    c. graphic
    d. mathematical

13. What is the relationship between the distance on a map and the actual distance on the Earth called?
    a. legend
    b. elevation
    c. relief
    d. scale

14. The latitude of the North Pole is
    a. 100° north. c. 180° north.
    b. 90° north. d. 90° south.

15. Widely spaced contour lines indicate a
    a. steep slope.
    b. gentle slope.
    c. hill.
    d. river.

Concept Mapping Transparency 2

Blackline masters of this Chapter Review can be found in the **Study Guide.**

16. __?__ is the height of an object above sea level.
   a. Contour interval
   b. Elevation
   c. Declination
   d. Index contour

Short Answer

17. How can a magnetic compass be used to find direction on the Earth's surface?

18. Why is a map legend important?

19. Why does Greenland appear so large in relation to other landmasses on a map with a Mercator projection?

20. What is the function of contour lines on a topographic map?

Concept Mapping

21. Use the following terms to create a concept map: maps, legend, map projection, map parts, scale, cylinder, title, cone, plane, date, north arrow.

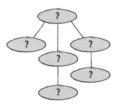

CRITICAL THINKING AND PROBLEM SOLVING

Write one or two sentences to answer the following questions:

22. One of the important parts of a map is its date. Why is this so important?

23. A mapmaker has to draw one map for three different countries that do not share a common unit of measure. What type of scale would this mapmaker use? Why?

24. How would a topographic map of the Rocky Mountains differ from a topographic map of the Great Plains?

MATH IN SCIENCE

25. A map has a verbal scale of 1 cm equals 200 m. If the actual distance between two points is 12,000 m, how far apart will they appear on the map?

26. On a topographic map, the contour interval is 50 ft. The bottom of a mountain begins on a contour line marked with a value of 1050 ft. The top of the mountain is within a contour line that is 12 lines higher than the bottom of the mountain. What is the elevation of the top of the mountain?

INTERPRETING GRAPHICS

Use the topographic map below to answer the questions that follow.

27. What is the elevation change between two adjacent lines on this map?

28. What type of relief does this area have?

29. What surface features are shown on this map?

30. What is the elevation at the top of Ore Hill?

Reading Check-up

Take a minute to review your answers to the Pre-Reading Questions found at the bottom of page 2. Have your answers changed? If necessary, revise your answers based on what you have learned since you began this chapter.

25

Short Answer

17. The needle on a magnetic compass points to magnetic north, indicating the direction of the magnetic north pole. If you know an area's magnetic declination, you can determine true north.

18. A map legend is important because it defines the set of symbols used in the map.

19. Greenland appears large on a map with a Mercator projection because of distortion. Maps with a Mercator projection are increasingly distorted as the distance from the equator increases.

20. Contour lines on a topographic map show the elevation, the relief, and the shape of landforms.

Concept Mapping

21.

An answer to this exercise can be found at the front of this book.

CRITICAL THINKING AND PROBLEM SOLVING

22. A date on a map is important because the Earth is constantly changing. The date shows you how old the information is. It might also indicate what the landscape was once like.

23. The mapmaker should use a representative fraction. A representative fraction is unitless and could be used by all three countries. Furthermore, the countries could find out distances on the map using their own unit of measure.

24. A topographic map of the Rocky Mountains would show contour lines close together, indicating steep slopes, while a contour map of the Great Plains would show contour lines spaced far apart, indicating a flat or gradual slope.

MATH IN SCIENCE

25. 60 cm
26. above 1,650 ft

INTERPRETING GRAPHICS

27. 20 ft
28. It has very high relief.
29. Answers will vary. Sample answer: Two hills are shown on this map.
30. The elevation at the top of Ore Hill is 2,025 ft.

Background

According to legend, Allah became displeased with the wickedness of the citizens of Ubar and buried the city under a wave of sand. Ubar remained lost for thousands of years until the coordinated efforts of filmmaker Nicholas Clapp, NASA scientist Dr. Ronald Blom, and a team of explorers uncovered the ruins in 1991.

Science, Technology, and Society

The Lost City of Ubar

Can you imagine tree sap being more valuable than gold? Well, about 2,000 years ago, a tree sap called frankincense was just that! Frankincense was used to treat illnesses and to disguise body odor. Ancient civilizations from Rome to India treasured it. While the name of the city that was the center of frankincense production and export had been known for generations—Ubar—there was just one problem: No one knew where it was! But now the mystery is solved. Using remote sensing, scientists have found clues hidden beneath desert sand dunes.

▲ *Trails and roads appear as purple lines on this computer-generated remote-sensing image.*

Using Eyes in the Sky

The process of remote sensing uses satellites to take pictures of large areas of land. The satellite records images as sets of data and sends these data to a receiver on Earth. A computer processes the data and displays the images. These remote sensing images can then be used to reveal differences unseen by the naked eye.

Remote-sensing images reveal modern roads as well as ancient caravan routes hidden beneath sand dunes in the Sahara Desert. But how could researchers tell the difference between the two? Everything on Earth reflects or radiates energy. Soil, vegetation, cities, and roads all emit a unique wavelength of energy. The problem is, sometimes modern roads and ancient roads are difficult to distinguish. The differences between similar objects can be enhanced by assigning color to an area and then displaying the area on a computer screen. Researchers used differences in color to distinguish between the roads of Ubar and modern roads. When researchers found ancient caravan routes and discovered that all the routes met at one location, they knew they had discovered the lost city of Ubar!

Continuing Discovery

Archaeologists continue to investigate the region around Ubar. They believe the great city may have collapsed into a limestone cavern beneath its foundation. Researchers are continuing to use remote sensing to study more images for clues to aid their investigation.

Think About It!

▶ Do modern civilizations value certain products or resources enough to establish elaborate trade routes for their transport? If so, what makes these products so valuable? Record your thoughts in your ScienceLog.

26

Answer to Think About It!

Students' answers will vary but may include the value modern civilizations place on paper currency, fossil fuels, or spices.

CAREERS

WATERSHED PLANNER

Have you ever wondered if the water you drink is safe or what you could do to make sure it stays safe? As a watershed planner, **Nancy Charbeneau** identifies and solves land-use problems that may affect water quality.

Nancy Charbeneau enjoys using her teaching, biology, and landscape architecture background in her current career as a watershed planner. A watershed is any area of land where water drains into a stream, river, lake, or ocean. Charbeneau spends a lot of time writing publications and developing programs that explain the effects of land use on the quality of water.

Land is used in hundreds of ways. Some of these land uses can have negative effects on water resources. Charbeneau produces educational materials to inform the public about threats to water quality.

Mapping the Problems

Charbeneau uses Geographic Information System (GIS) maps to determine types of vegetation and the functions of different sections of land. GIS is a computer-based tool that allows people to store, access, and display geographic information collected through remote-sensing, field work, global positioning systems, and other sources. Maps and mapping systems play an important role in identifying land areas with water problems.

Maps tell Charbeneau whether an area has problems with soil erosion that could threaten water quality. Often the type of soil plays an important role in erosion. Thin or sandy soil does not hold water well, allowing for faster runoff and erosion. Flat land with heavy vegetation holds more water and is less prone to erosion.

Understanding the Importance

Charbeneau's biggest challenge is increasing understanding of the link between land use and water quality. If a harmful substance is introduced into a watershed, it may contaminate an aquifer or a well. As Charbeneau puts it, "Most people want to do the right thing, but they need information about land management practices that will protect water quality but still allow them to earn a decent living off their land."

Reading the Possibilities

▶ Map out your nearest watershed. Can you find any potential sources of contamination?

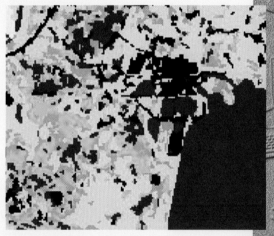

▲ *This GIS map shows the location of water in blue.*

27

Background

Nancy Charbeneau has undergraduate degrees in elementary education and biology and a master's degree in landscape architecture.

Charbeneau's experience as a teacher prepared her for the production of educational materials that alert the public to water quality issues. Her degree in landscape architecture exposed her to the sophisticated mapping systems she uses as a watershed developer.

Answer to Reading the Possibilities

Answers will vary, depending on the topography and degree of urbanization of your region. Industrialized areas may have problems with groundwater pollution, while mountainous areas may have problems with erosion. Aerial photography, GIS, and infrared maps can reveal additional information about the vegetation and land use in a region.

Chapter Organizer

CHAPTER ORGANIZATION	TIME MINUTES	OBJECTIVES	LABS, INVESTIGATIONS, AND DEMONSTRATIONS
Chapter Opener pp. 28–29	45	National Standards: UCP 2, SAI 1, SPSP 5, ES 2a	**Start-Up Activity,** What's the Difference? p. 29
Section 1 Weathering	45	▶ Describe how ice, rivers, tree roots, and animals cause mechanical weathering. ▶ Describe how water, acids, and air cause chemical weathering of rocks. SAI 1, ES 1c, 1k; Labs UCP 2, SAI 1	**Making Models,** Rockin' Through Time, p. 46 **Datasheets for LabBook,** Rockin' Through Time **Discovery Lab,** Great Ice Escape, p. 94 **Datasheets for LabBook,** Great Ice Escape **QuickLab,** Acids React! p. 34 **Whiz-Bang Demonstrations,** When It Rains, It Fizzes
Section 2 Rates of Weathering	90	▶ Explain how the composition of rock affects the rate of weathering. ▶ Describe how a rock's total surface area affects the rate at which it weathers. ▶ Describe how mechanical and chemical weathering work together to break down rocks and minerals. ▶ Describe how differences in elevation and climate affect the rate of weathering. SAI 1, ES 1c	**EcoLabs and Field Activities,** Whether It Weathers (or Not)
Section 3 From Bedrock to Soil	90	▶ Define *soil.* ▶ Explain the difference between residual and transported soils. ▶ Describe the three soil horizons. ▶ Describe how various climates affect soil. ES 1e, 1k	**Calculator-Based Labs,** A Soil Study
Section 4 Soil Conservation	90	▶ Describe three important benefits that soil provides. ▶ Describe three methods of preventing soil erosion. SPSP 2, 4, 5	**Long-Term Projects & Research Ideas,** Precious Soil

See page **T23** *for a complete correlation of this book with the*

NATIONAL SCIENCE EDUCATION STANDARDS.

TECHNOLOGY RESOURCES

 Guided Reading Audio CD English or Spanish, Chapter 2

 One-Stop Planner CD-ROM with Test Generator

 CNN. Eye on the Environment, Prairie Restoration, Segment 16

Science, Technology & Society, Homemade Dirt, Segment 16

CLASSROOM WORKSHEETS, TRANSPARENCIES, AND RESOURCES	SCIENCE INTEGRATION AND CONNECTIONS	REVIEW AND ASSESSMENT
Directed Reading Worksheet **Science Puzzlers, Twisters & Teasers**		
Directed Reading Worksheet, Section 1 **Transparency 138,** Chemical Weathering **Reinforcement Worksheet,** Autobiography of a Rock **Transparency 259,** pH Values of Common Materials	**Chemistry Connection,** p. 30 **Multicultural Connection,** p. 30 in ATE **Connect to Physical Science,** p. 31 in ATE **Connect to Life Science,** p. 32 in ATE **Connect to Physical Science,** p. 33 in ATE	**Self-Check,** p. 32 **Homework,** p. 32 in ATE **Section Review,** p. 35 **Quiz,** p. 35 in ATE **Alternative Assessment,** p. 35 in ATE
Directed Reading Worksheet, Section 2 **Transparency 139,** Weathering and Surface Area	**Connect to Physical Science,** p. 36 in ATE **MathBreak,** The Power of 2, p. 37	**Section Review,** p. 38 **Quiz,** p. 38 in ATE **Alternative Assessment,** p. 38 in ATE
Directed Reading Worksheet, Section 3 **Transparency 140,** Soil Horizons	**Connect to Life Science,** p. 40 in ATE **Multicultural Connection,** p. 41 in ATE **Connect to Environmental Science,** p. 41 in ATE **Across the Sciences:** Worms of the Earth, p. 52	**Homework,** p. 39 in ATE **Section Review,** p. 42 **Quiz,** p. 42 in ATE **Alternative Assessment,** p. 42 in ATE
Directed Reading Worksheet, Section 4 **Critical Thinking Worksheet,** Buying the Farm **Reinforcement Worksheet,** Where the Tall Corn Grows	**Connect to Life Science,** p. 43 in ATE **Multicultural Connection,** p. 44 in ATE **Math and More,** p. 45 in ATE **Eye on the Environment:** Losing Ground, p. 53	**Homework,** p. 44 in ATE **Section Review,** p. 45 **Quiz,** p. 45 in ATE **Alternative Assessment,** p. 45 in ATE

 internet **connect**

 go.
hrw
.com **Holt, Rinehart and Winston**
On-line Resources
go.hrw.com

For worksheets and other teaching aids related to this chapter, visit the HRW Web site and type in the keyword: **HSTWSF**

 SC/LINKS
NSTA
National Science
Teachers Association
www.scilinks.org

Encourage students to use the *sci*LINKS numbers listed in the internet connect boxes to access information and resources on the **NSTA** Web site.

END-OF-CHAPTER REVIEW AND ASSESSMENT

Chapter Review in Study Guide
Vocabulary and Notes in Study Guide
Chapter Tests with Performance-Based Assessment, Chapter 2 Test
Chapter Tests with Performance-Based Assessment, Performance-Based Assessment 2
Concept Mapping Transparency 10

Chapter Resources & Worksheets

Visual Resources

TEACHING TRANSPARENCIES

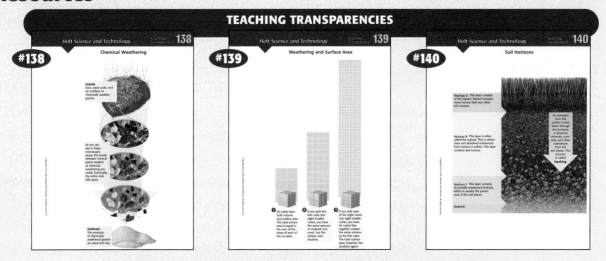

#138 — Chemical Weathering

#139 — Weathering and Surface Area

#140 — Soil Horizons

TEACHING TRANSPARENCIES

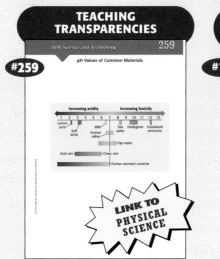

#259 — pH Values of Common Materials

LINK TO PHYSICAL SCIENCE

CONCEPT MAPPING TRANSPARENCY

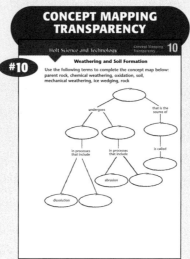

#10 — Weathering and Soil Formation

Use the following terms to complete the concept map below: parent rock, chemical weathering, oxidation, soil, mechanical weathering, ice wedging, rock

Meeting Individual Needs

DIRECTED READING

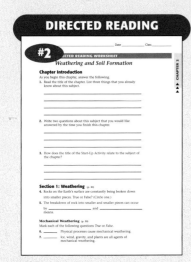

#2 — DIRECTED READING WORKSHEET
Weathering and Soil Formation

REINFORCEMENT & VOCABULARY REVIEW

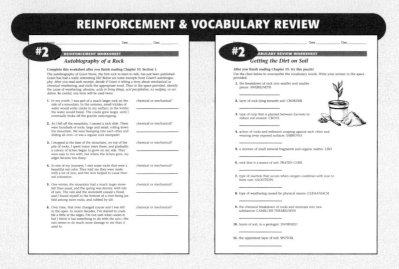

#2 — REINFORCEMENT WORKSHEET
Autobiography of a Rock

#2 — VOCABULARY REVIEW WORKSHEET
Getting the Dirt on Soil

SCIENCE PUZZLERS, TWISTERS & TEASERS

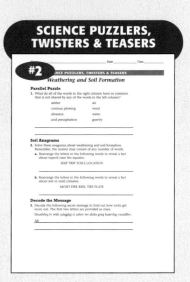

#2 — SCIENCE PUZZLERS, TWISTERS & TEASERS
Weathering and Soil Formation

Chapter 2 • Weathering and Soil Formation

Review & Assessment

STUDY GUIDE

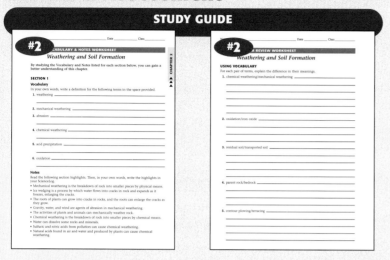

CHAPTER TESTS WITH PERFORMANCE-BASED ASSESSMENT

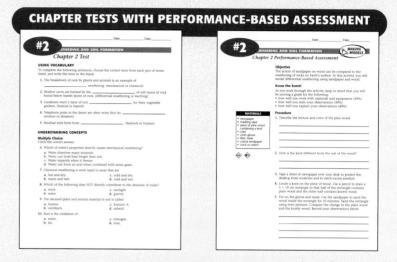

Lab Worksheets

WHIZ-BANG DEMONSTRATIONS

ECOLABS & FIELD ACTIVITIES

LONG-TERM PROJECTS & RESEARCH IDEAS

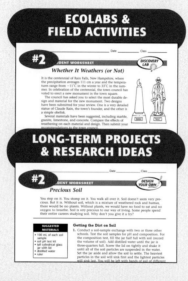

DATASHEETS FOR LABBOOK

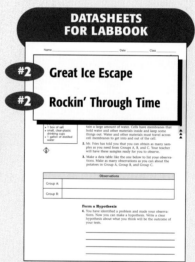

Applications & Extensions

CRITICAL THINKING & PROBLEM SOLVING

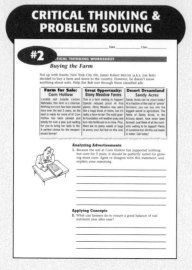

EYE ON THE ENVIRONMENT

SCIENCE TECHNOLOGY

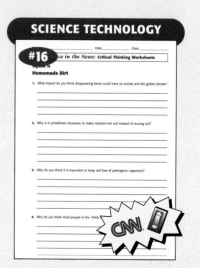

SECTION 1

Weathering

▶ Thermal Contraction and Expansion

There is much scientific debate over whether the daily and seasonal heating and cooling of rocks causes wide-scale weathering. In desert environments, where temperature ranges can be extreme, small rocks can shatter from expansion and contraction. But does this type of weathering occur in larger rocks and in climates that are more temperate? Geologists attempting to replicate this process in a lab have had little success. In one experiment granite samples were repeatedly heated and cooled by more than 100°C, and no fracturing was observed. This suggests that if thermal expansion and contraction weathers rock, it may do so over the course of hundreds of thousands of years.

IS THAT A FACT!

➤ Before the invention of power drills and saws, stonemasons sometimes filled existing joints and cracks in rocks with water and waited for ice wedging to split the rocks. Obviously, this method was effective only in areas where temperatures dropped to freezing or below.

▶ Salt Cracking

In places where ground water contains dissolved salts, salt water seeps into bedrock. When the water evaporates, the dissolved salts crystallize, and the growing crystals can exert enough force to fracture rock. This process, known as salt cracking, can be seen at the ocean, where sea cliffs become pitted and cracked from salt deposits. In desert regions salt cracking erodes the base of some sandstone formations, leaving an unweathered rock balancing on an eroded pedestal.

▶ Sandstorms

During a sandstorm, the wind may blow more than 160 km/h. The friction created by moving sand grains not only abrades rock but also generates static electricity. Some people caught in sandstorms have experienced painful headaches caused by the buildup of static electricity.

SECTION 2

Rates of Weathering

▶ Mineral Composition and Weathering Rates

The order in which minerals crystallize from magma is nearly the same as the order in which they weather. Minerals that form quickly and at high temperatures and pressures within Earth tend to be unstable at the surface, and they are less resistant to chemical weathering. Minerals that form slowly and at lower temperatures are much more resistant to the effects of weathering.

From Bedrock to Soil

▶ Types of Soils in the United States

The soils of the mainland United States can be divided into two major types—*pedocal* and *pedalfer.* Pedocal is a calcium-rich soil that covers most of the western United States. Pedocal gets its name from the Latin *ped,* meaning "soil," joined with *cal,* representing calcium. Pedalfer is an iron- and aluminum-rich soil that covers most of the eastern half of the country. The *al* in pedalfer stands for aluminum; the *fer* stands for ferrum (iron).

▶ Salinization

All ground water contains small concentrations of salts. If arid or semiarid soils are intensively irrigated, the soil can accumulate so much salt that it cannot support plant life. This process, called salinization, can ruin croplands. Some historical scholars argue that salinization contributed to the decline and fall of many ancient societies, including the Babylonian civilization.

IS THAT A FACT!

- ▶ *Regolith* is a term that describes all of the weathered material that lies over the bedrock. *Soil* refers to the upper layers of the regolith that can support plant life.

- ▶ The term *regolith* is derived from the Greek word *rhegos,* meaning "blanket," and the Old English *lithos* meaning "rock." This derivation is important because it denotes the protective qualities of soil. Like a blanket, the soil protects the rock below from weathering. In mountain regions where soil is easily eroded, bedrock weathers much more quickly.

Soil Conservation

▶ Farming in the Imperial Valley

Although desert soils are low in organic matter, they are not necessarily poor soils. Desert soil such as that of the Imperial Valley in California is actually quite rich with the minerals needed for plant growth. Water diverted from the Colorado River is used to irrigate the valley, and it is now one of the nation's major farming regions, growing such crops as alfalfa, cotton, and sugar beets. While the Imperial Valley is incredibly productive, agriculture in the region relies heavily on the use of fertilizers. There is also much debate over whether the Imperial Valley diverts too much water from the Colorado River. In addition, people are concerned because irrigation has concentrated salts in the soil and polluted the nearby Salton Sea.

▶ Federal Soil Conservation Service

In response to the devastating windstorms that swept across the Great Plains, the U.S. Department of Agriculture formed the Soil Conservation Service in 1935. Working with ranchers and farmers, conservationists instituted such strategies as contour plowing and terracing, planting trees as windbreaks, allowing land to lie fallow, and planting drought-resistant crops.

IS THAT A FACT!

- ▶ In some places across the United States, soil conservation researchers have implemented an innovative approach to roadside erosion control. Organic wastes, which otherwise would be dumped in landfills, are spread along the sides of highways. The nutrient-rich wastes help grasses grow, and the grass roots anchor soil in place.

For background information about teaching strategies and issues, refer to the *Professional Reference for Teachers.*

Pre-Reading Questions

Students may not know the answers to these questions before reading the chapter, so accept any reasonable response.

Suggested Answers

1. Answers will vary. Both wind and water cause the abrasion of rock surfaces. Students may also mention ice wedging or chemical weathering.

2. Answers will vary. Soil supports plant life and plants are the base of the food chain.

3. Answers will vary. Weathering breaks down rocks, contributing to the formation of soil.

Sections

Pre-Reading Questions

1. How do water and air cause rocks to crumble?

2. Why is soil one of our most important resources?

3. How are weathering and soil formation related?

28

Weathering and Soil Formation

NICE AND COZY

The American badger *(Taxidea taxus)* is the only badger living in the Western Hemisphere. It is most often found in dry, open areas of western North America. Although the badger stands only about 23 cm high, it is a muscular animal that can dig very rapidly. Badgers are known for their burrowing ability, and they dig in the soil for food and to build underground homes. In this chapter, you will learn about ways that animals like the badger use and help create soil.

---

internetconnect

HRW On-line Resources

go.hrw.com

For worksheets and other teaching aids, visit the HRW Web site and type in the keyword: **HSTWSF**

www.scilinks.com

Use the *sci*LINKS numbers at the end of each chapter for additional resources on the **NSTA** Web site.

Smithsonian Institution®

www.si.edu/hrw

Visit the Smithsonian Institution Web site for related on-line resources.

www.cnnfyi.com

Visit the CNN Web site for current events coverage and classroom resources.

WHAT'S THE DIFFERENCE?

In this chapter, you will learn about the processes and rates of weathering. Do this activity to learn about how the size and surface area of a material affect how quickly the substance breaks down.

Procedure

1. Fill **two small containers** about half full with **water.**

2. Add **one sugar cube** to one container.

3. Add 1 tsp of **granulated sugar** to the other container.

4. Using **two different spoons,** stir the water and sugar in each container at the same rate.

5. Using a **stopwatch,** measure how long it takes for the sugar to dissolve in each container.

Analysis

6. Did the sugar dissolve at the same rate in both containers? Explain why or why not.

7. Which do you think would wear away faster—a large rock or a small rock? Explain your answer.

WHAT'S THE DIFFERENCE?

MATERIALS

FOR EACH GROUP:
- 2 small containers
- water
- sugar cube
- 1 tsp. granulated sugar
- 2 different spoons
- stopwatch or timepiece with second hand

Answers to START-UP Activity

6. The granulated sugar dissolved faster than the sugar cube because the grains of sugar had more surface area than the sugar cube. Therefore, more of the granulated sugar dissolved in a shorter period of time.

7. A small rock would wear away faster because it has more surface area for weathering to affect.

Focus

Weathering

In this section, students will learn how mechanical processes such as ice wedging, abrasion, and plant and animal activities contribute to the weathering of rock. Students will also learn how water, acid precipitation, oxidation, and the growth of lichens cause chemical weathering of rock.

 **Bellringer**

Ask students to think about how potholes form in paved roads. Have them write a few sentences in their ScienceLog that describe how water contributes to the formation of potholes. Students should illustrate how cycles of freezing and thawing cause potholes to grow. **Sheltered English**

① Motivate

GROUP ACTIVITY

Have groups of students find photographs in magazines that illustrate weathering. Examples might include rusted cars, mailboxes or bikes; sidewalks or walls that have been cracked by plant roots; potholes; and weathered statues and buildings. Then have groups try to find photographs of rock formations that depict similar kinds of weathering. Ask the class to help you group the photographs into examples of physical weathering and chemical weathering.

 Directed Reading Worksheet Section 1

Terms to Learn

weathering
mechanical weathering
abrasion
chemical weathering
acid precipitation
oxidation

What You'll Do

◆ Describe how ice, rivers, tree roots, and animals cause mechanical weathering.
◆ Describe how water, acids, and air cause chemical weathering of rocks.

Chemistry CONNECTION

Almost all liquids contract when they freeze to form a solid—their volume decreases and their density increases. When these substances freeze, the frozen solid sinks. Just the opposite occurs to water when it freezes. Water expands and becomes less dense, which is why ice floats in water.

30

Weathering

Weathering is the breakdown of rock into smaller and smaller pieces. Rocks on Earth's surface are undergoing weathering all the time, either by mechanical means or by chemical means. You will learn the difference as you read on. You will also learn how these processes shape the surface of our planet.

Mechanical Weathering

If you were to crush one rock with another rock, you would be demonstrating one type of mechanical weathering. **Mechanical weathering** is simply the breakdown of rock into smaller pieces by physical means. Agents of mechanical weathering include ice, wind, water, gravity, plants, and even animals.

Ice As you know, water has the unusual property of expanding when it freezes. (This is just the opposite of most substances.) When water seeps into a crack in a rock during warm weather and then freezes during cold weather, it expands. And when it expands, it pushes against the sides of the crack, forcing it to open wider. This process is called *ice wedging*. **Figure 1** shows how ice wedging occurs over time.

Figure 1 *The granite at right has been broken down by repeated ice wedging, as shown in the illustration below.*

Water   Ice   Water   Ice

 **Multicultural CONNECTION**

In the second century B.C., Buddhist monks began carving an intricate system of caves in a massive basalt flow in central India. The Ajanta caves comprised a complex of monasteries, temples, and living quarters. The caves were adorned with beautiful frescoes and carvings and then mysteriously abandoned in the seventh century. They were rediscovered by British game hunters less than 200 years ago. The Ajanta caves are notable not only for their artwork but also for the manner in which they were carved. The monks first cut channels in the rock and then jammed dry logs in the crevices. They poured water on top of the logs and waited for the expanding wood to shatter the rock. In this way, they carved 30 caves out of solid rock.

Wind, Water, and Gravity When you write on a chalkboard, a process called *abrasion* takes place. As you scrape the piece of chalk against the chalkboard, some of the chalk rubs off to make a line on the board. As particles of chalk are worn off, the piece of chalk wears down and becomes more rounded at the tip. The same thing happens to rocks. In nature, **abrasion** is the action of rocks and sediment grinding against each other and wearing away exposed surfaces.

Abrasion can happen in many ways. For example, when rocks and pebbles roll along the bottom of swiftly flowing rivers, they bump into and scrape against each other. They eventually become river rocks, as shown in **Figure 2**.

Wind also causes abrasion. For example, when wind blows sand against exposed rock, the sand eventually wears away the rock's surface. **Figure 3** shows what this kind of sandblasting can do.

Abrasion also occurs when rocks fall on one another. **Figure 4** shows a rock slide. You can imagine the forces rocks exert on each other as they tumble down a mountainside. In fact, any time one rock hits another, abrasion takes place.

Figure 2 *These river rocks are rounded because they have been tumbled in the riverbed by fast-moving water for many years.*

Figure 3 *This rock has been shaped by blowing sand. Such rocks are called* ventifacts.

Figure 4 *Rocks grind against each other in a rock slide, creating smaller and smaller rock fragments.*

31

IS THAT A FACT!

Mount Fuji is a dormant volcano that is a source of national pride among the Japanese. Unfortunately, the forces of mechanical weathering threaten to change the volcano's conical shape and near-perfect symmetry. To preserve the mountain's shape, the Japanese government has built a 17 m long concrete brace over a widening crevice near the mountain's summit. Before action was taken, as much as 300,000 tons of rock and soil had fallen down the mountainside every year.

READING STRATEGY

Prediction Guide After students read the definition of *weathering* in the first paragraph of this section, have them decide whether the following statements are true or false:

- Very hard rocks do not weather. (false)
- Water can weather rock. (true)
- A snowflake can be acidic enough to weather rock. (true)
- Plants play an important role in weathering rock. (true)

CONNECT TO
PHYSICAL SCIENCE

Another process of mechanical weathering is called *exfoliation*. Rocks generally form under great pressure. As the material overlying a rock formation is removed by erosion and tectonic activity, the pressure on the formation is reduced. As the pressure is reduced, the rock expands in volume, and long, curved cracks develop parallel to the rock's surface. In this way, an outcrop "sheds" layers of rock. Exfoliation can often be observed in granite outcrops. Ask students to describe why granite formations are prone to exfoliation. (Granite forms underground, so it forms under a great deal of pressure from the rock above. A granite pluton 15 km underground forms at 5,000 times the pressure at Earth's surface. As the granite is pushed toward the surface, and the overlaying rock is weathered away, the pressure on the rock is reduced, and the granite exfoliates.)

LabBook **PG 94**
The Great Ice Escape

MISCONCEPTION ALERT

Students may be surprised to learn that animals such as earthworms, coyotes, and rabbits play significant roles in weathering rock. Point out that human activity also contributes to the weathering of rock and to the formation of soil. People move large amounts of soil and rock whenever they farm, build, or drive off-road vehicles. In addition, people blast through rock to make tunnels, roads, mines, and quarries. If you also consider the effect of air pollution on the weathering of rock, the effects of human activity become even more significant.

CONNECT TO LIFE SCIENCE

As ground-dwelling termites construct their homes, they excavate an enormous amount of soil and rock fragments. In some parts of the world, termites must burrow to great depths to mine the wet clay they need to build their mounds. Occasionally, the termites strike it rich. Geochemical prospectors have learned from indigenous cultures in Africa, Asia, Australia, and South America to analyze termite mounds for ore deposits such as tin, silver, gold, diamonds, and uranium. In some parts of Africa, gold concentrations in termite mounds are rich enough that people earn money by panning gold from them.

Figure 5 *Although they grow slowly, tree roots are strong enough to break solid rock.*

Self-Check

Describe the property of water that causes ice wedging. *(See page 120 to check your answer.)*

Figure 6 *Animals that live in the soil cause a lot of weathering.*

Plants You may not think of plants as being strong, but some plants can easily break rocks. Have you ever seen how tree roots can crack sidewalks and streets? Roots aren't fast, but they certainly are powerful! Plants often send their roots into existing cracks in rocks. As the plant gets bigger, the force of the expanding root becomes so strong that the crack is made larger. Eventually, the entire rock can split apart, as you can see in **Figure 5.**

Animals Believe it or not, earthworms cause a lot of weathering! They burrow through the soil and move soil particles around. This exposes fresh surfaces to continued weathering. Would you believe that some kinds of tropical worms move an estimated 100 metric tons of soil per acre every year? Almost any animal that burrows causes mechanical weathering. Ants, mice, coyotes, and rabbits all make their contribution. **Figure 6** shows some of these animals in action. The mixing and digging that animals do often contribute to another type of weathering, called *chemical weathering.* You will learn about this next.

Answer to Self-Check

Water expands as it freezes. This expansion exerts a force great enough to crack rock.

Homework

Making Observations Ask students to keep a journal in which they note processes of weathering and soil formation in your area. Ask them to pay particular attention to plant and animal activity. Encourage students to write a short entry each day detailing their observations.

Chemical Weathering

If you place a drop of strong acid on a rock, it will probably "eat away" a small part of the rock. This is an example of chemical weathering. **Chemical weathering** is the chemical breakdown of rocks and minerals into new substances. The most common agents of chemical weathering are water, weak acids, air, and soil. **Figure 7** shows the chemical weathering of granite.

Water If you drop a sugar cube into a glass of water, it will dissolve after a few minutes. If you drop a piece of chalk into a glass of water, it will also dissolve, only much slower than a sugar cube. Both cases are examples of chemical weathering. Even hard rock, like granite, is broken down by water; it just may take a few thousand years.

Acid Precipitation A car battery contains sulfuric acid, a very dangerous acid that should never come in contact with your skin. You may already know that the concentrated form of sulfuric acid found in batteries does not occur naturally on Earth. However, a weaker form of sulfuric acid does exist in nature. In fact, precipitation such as rain and snow naturally contains small amounts of sulfuric and nitric acid. These acids can slowly break down rocks and other materials that they come in contact with.

Precipitation that contains acids due to air pollution is called **acid precipitation.** Acid precipitation contains more acid than normal precipitation, so it can cause very rapid weathering of rock. Even the bronze statue shown in **Figure 8** is being chemically weathered by acid precipitation.

Figure 8 This statue is being damaged by acid precipitation.

Figure 7 After thousands of years of chemical weathering, even hard rock, like granite, can turn to sediment.

Granite
Rain, weak acids, and air combine to chemically weather granite.

As you can see in these microscopic views, the bonds between mineral grains weaken as chemical weathering proceeds. Eventually, the entire rock falls apart.

Sediment
The products of chemically weathered granite are sand and clay.

33

Teaching Transparency 138
"Chemical Weathering"

Reinforcement Worksheet
"Autobiography of a Rock"

COOPERATIVE LEARNING

Have student groups focus on different aspects of acid precipitation, and have the class organize a task force presentation for your school or community.

- The **monitoring group** can test the pH of precipitation in your area by using litmus paper. Students can also test the pH of tap water, surface runoff, rivers, and lakes. Have students contact the local weather service to find records of the pH of precipitation in your area over several decades. The monitoring group can present its findings in graphs and other visual displays.

- The **research group** can prepare a presentation on the causes and effects of acid precipitation. They should also focus on legislation and other solutions for the air pollution problems that contribute to acid precipitation.

CONNECT TO
PHYSICAL SCIENCE

Students are often confused by the pH scale. The term *pH* is French and translates as "power of hydrogen." It refers to the concentration of hydrogen ions in a solution. In a measurement of the pH of an acid, a decrease of one number on the pH scale represents an increase in the concentration of hydronium ions by a power of ten. Thus an acid with a pH of 2 is one hundred times as concentrated as an acid with a pH of 4. Remind students that acidic solutions have a pH less than 7, and that basic solutions have a pH greater than 7.

Teaching Transparency 259
"pH Values of Common Materials"

LINK TO PHYSICAL SCIENCE

3 Extend

 QuickLab

Safety Caution: Students with allergies to tomatoes should use a cotton swab to apply the ketchup.

Answers to QuickLab

2. Answers will vary. Students might note that the grime reacted chemically with the acid in the ketchup and dissolved.

3. Answers will vary. Students should note that rocks react with acids in the same way that the grime reacted with the ketchup.

RETEACHING

To reinforce the difference between chemical and mechanical weathering, show students two matches. To demonstrate mechanical weathering, break one of the matches. To show chemical weathering, light the other match and let it burn. Next, have students decide whether each of the following phenomena is an example of mechanical or chemical weathering.

- a rockfall on a mountainside (mechanical)
- a rusty bridge (chemical)
- lichens and mosses growing on a boulder (chemical)
- an alpine glacier advancing down a valley (mechanical)
- a mole burrowing in the ground (mechanical)
  Sheltered English

COOPERATIVE LEARNING

Have groups work together to write a five-question quiz about the material in Section 1. Have the groups exchange quizzes and work together to answer the questions.

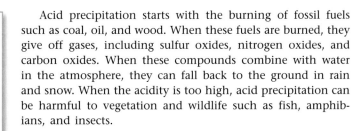

Acids React!

Have you ever heard someone refer to a certain food as being "acidic"? You consume acids in your food every day. For example, ketchup contains weak acids that can react with certain substances in a rather dramatic way. Try this:

1. Take a **penny** that has a dull appearance, rub **ketchup** on it for several minutes, and then rinse it off.

2. Where did all the grime go?

3. How is this similar to what happens to a rock when it is exposed to natural acids during weathering?

TRY at HOME

Figure 9 *Papoose Room, Carlsbad Caverns. Carlsbad Caverns National Park, New Mexico.*

Acid precipitation starts with the burning of fossil fuels such as coal, oil, and wood. When these fuels are burned, they give off gases, including sulfur oxides, nitrogen oxides, and carbon oxides. When these compounds combine with water in the atmosphere, they can fall back to the ground in rain and snow. When the acidity is too high, acid precipitation can be harmful to vegetation and wildlife such as fish, amphibians, and insects.

Acid in Ground Water In certain places ground water contains weak acids, such as carbonic or sulfuric acid. When this ground water comes in contact with limestone, the limestone breaks down. Over a long period of time, this can have some spectacular results. Enormous caverns, like the one shown in **Figure 9,** can form as the limestone is eaten away. Limestone, you may remember, is made of calcite, which reacts strongly with acid.

Acids in Living Things Another source of acids for weathering might surprise you. Take a look at **Figure 10.** Lichens produce organic acids that can slowly break down rock. If you have ever taken a walk in a park or forest, you have probably seen lichens growing on the sides of trees or rocks. Lichens can also grow in places where some of the hardiest plants cannot. Lichens can be found in deserts, in arctic areas, and in areas high above timberline, where even trees don't grow.

Figure 10 *Lichens, which consist of fungi and algae living together, contribute to chemical weathering.*

🔗 **internetconnect**

SC*i*LINKS
NSTA

TOPIC: Weathering
GO TO: www.scilinks.org
***sci*LINKS NUMBER:** HSTE230

Air The car shown in **Figure 11** is undergoing chemical weathering due to the air. The oxygen in the air is reacting with the iron in the car, causing the car to rust. Water speeds up the process, but the iron would rust even if no water were present. This process also happens in certain types of rocks, particularly those containing iron, as you can see in **Figure 12.** Scientists call this process *oxidation*.

Oxidation is a chemical reaction in which an element, such as iron, combines with oxygen to form an oxide. (The chemical name for rust is *iron oxide*.) Oxidation is a common type of chemical weathering, and rust is probably the most familiar result of oxidation.

— **Activity** —

Imagine that you are a tin can—shiny, new, and clean. But something happens, and you don't make it to a recycling bin. Instead, you are left outside at the mercy of the elements. In light of what you have learned about physical and chemical weathering, write a story about what happens to you over a long period of time. What is your ultimate fate?

TRY at HOME

Figure 12 *The red color of the rock at Capitol Reef National Park is due to the oxidation of iron.*

Figure 11 *Rust is a result of chemical weathering.*

SECTION REVIEW

1. Describe three ways abrasion occurs in nature.

2. Describe the similarity between the ways tree roots and ice mechanically weather rock.

3. **Making Generalizations** Why does acid precipitation weather rocks faster than normal rain does?

internet**connect**

SCI**LINKS**
NSTA

TOPIC: Weathering
GO TO: www.scilinks.org
*sci***LINKS NUMBER:** HSTE230

BRAIN FOOD

Studying rock strata for signs of the oxidation of ferric minerals is like reading the rings of a tree: geologists can determine the concentration of oxygen in Earth's atmosphere throughout geologic time. Until about 3 billion years ago, the concentration of oxygen in Earth's atmosphere was not significant enough to oxidize iron. After that time, sedimentary deposits show layers of oxidized and unoxidized iron, indicating that oxygen concentrations fluctuated. Rocks formed after 1.8 billion years ago are uniformly oxidized. What does this suggest? (that oxygen concentrations have not changed significantly since that time)

4 Close

Quiz

1. How do earthworms aid in weathering? (When earthworms burrow, they move soil particles around. This exposes fresh surfaces to weathering.)

2. What human activities can increase the acidity of precipitation? (any activities that burn fossil fuels, such as driving cars, heating homes, and producing electricity)

ALTERNATIVE ASSESSMENT

Have groups do research to come up with five multiple-choice trivia cards for a game that tests the player's knowledge of weathering. Students can use the cards in a class trivia challenge.

▼ *Answers to Section Review*

1. Answers will vary. Sample answer: Abrasion can occur when rocks scrape against each other in a landslide, when rocks are tumbled by the water in a river, or when rocks are "sandblasted" by small particles carried in the wind.

2. Both tree roots and ice enter cracks in rock and expand, making the cracks wider and shattering the rock.

3. Acid precipitation is more acidic than regular rain, so it reacts faster with the minerals in rocks.

SECTION 2
READING WARM-UP

Focus

Rates of Weathering

This section explores why different types of rock weather at different rates, and how the surface area of rock affects the rate at which it weathers. In addition, students learn why elevation and climate are important factors affecting weathering rates.

Bellringer

Ask students to imagine that they are in a sand-castle-building competition at the beach. Ask them to come up with a variety of ways to protect their castle against the weathering effects of the wind and waves. Students can illustrate their ideas and share them with the class.

1 Motivate

GROUP ACTIVITY

Supply each group with one clear glass containing a calcium antacid tablet and a second containing a tablet cut into quarters. Tell students that antacid tablets and limestone both contain calcium carbonate, which dissolves in acidic solutions. Ask students to pour enough vinegar into the glasses to cover the tablets. As students observe the reactions, ask them which of the tablets "weathers" more rapidly. Students can repeat the experiment with a crushed tablet. Lead students to conclude that surface area affects the rate at which things weather. Ask students if the crushed tablet released more gas than the whole tablet. (No, the surface-area-to-volume ratio only affects the rate at which the reaction occurs.)

Terms to Learn

differential weathering

What You'll Do

- ◆ Explain how the composition of rock affects the rate of weathering.
- ◆ Describe how a rock's total surface area affects the rate at which it weathers.
- ◆ Describe how mechanical and chemical weathering work together to break down rocks and minerals.
- ◆ Describe how differences in elevation and climate affect the rate of weathering.

Rates of Weathering

Different types of rock weather at different rates. Some types of rock weather quickly, while other types weather slowly. The rate at which a rock weathers depends on many factors—climate, elevation, and, most important, what the rock is made of.

Differential Weathering

Hard rocks, such as granite, weather more slowly than softer rocks, such as limestone. This is because granite is made of minerals that are generally harder and more chemically stable than the minerals in limestone. **Differential weathering** is a process by which softer, less weather-resistant rocks wear away, leaving harder, more weather-resistant rocks behind.

Figure 13 shows a spectacular landform that has been shaped by differential weathering. Devils Tower, the core of an ancient volcano, was once a mass of molten rock deep within an active volcano. When the molten rock solidified, it was protected from weathering by the softer rock of the volcano. After thousands of years of weathering, the soft outer parts of the volcano have worn away, leaving the harder, more resistant rock of Devils Tower behind. Of course, not all landforms are this spectacular. But if you look closely, you can see the effects of differential weathering in almost any landscape.

Figure 13 *Devils Tower is a landform known as a* volcanic neck. *The illustration is an artist's conception of how the original volcano may have looked. The photo inset shows Devils Tower as it appears today.*

CONNECT TO
PHYSICAL SCIENCE

Why is the intrusive volcanic rock that makes up Devils Tower more resistant to weathering than the extrusive volcanic rock of the former volcano? Both rock types had the same composition. The difference is their *cohesiveness*. The intrusive rock cooled more slowly than the extrusive rock. As the rock slowly cooled, it formed large crystals that interlocked like a 3-D jigsaw puzzle. This made the volcanic neck more resistant to weathering. In contrast, the rock that made up the outside of the volcano cooled quickly. It was made of much smaller crystals and *groundmass* material—material that cooled so fast it didn't form crystals. Have students draw a series of time-lapse illustrations to show the formation of the volcanic neck in **Figure 13.**

The Shape of Weathering

As you know, weathering takes place on the outside surface of rocks. So the more surface area that is exposed to weathering, the faster the rock will be worn down. A large rock has a large surface area, but it also has a large volume. Because of this, it will take a long time for a large rock to wear down.

If a big rock is broken into smaller rocks, weathering occurs much more quickly. This is because a smaller rock has more surface area relative to its volume. This means that more of a small rock is exposed to the weathering process. The cubes in **Figure 14** show how this principle works.

Figure 14 *As surface area increases, total volume stays the same. Each square in the background represents the face of a cube.*

❶ All cubes have both volume and surface area. The total surface area is equal to the sum of the areas of each of the six sides.

❷ If you split the first cube into eight smaller cubes, you have the same amount of material (volume), but the surface area doubles.

❸ If you split each of the eight cubes into eight smaller cubes, you have 64 cubes that together contain the same volume as the first cube. The total surface area, however, has doubled again!

MATH BREAK

The Power of 2

You can calculate the surface area of a square or rectangle by multiplying its width times its length ($w \times l$). For example, one side of a cube that measures 5 cm by 5 cm has a surface area of 25 cm². Now you try:

What is the surface area of one side of a cube that is 8 cm wide and 8 cm long? What is the surface area of the entire cube? (Hint: A cube has six equal sides.)

37

2) Teach

COOPERATIVE LEARNING

Weathering Game Assign each student a number, and have the students stand in a square formation. Tell them that they are a rock formation exposed to the forces of weathering. Draw a number randomly. If the student assigned to that number is standing on the outside edge of the square, he or she "weathers" and gets to sit down. Students who are not "exposed" must remain standing. After all the numbers have been drawn, have students work as a class to come up with the fastest weathering square formation. Student groups will discover that squares that comprise four people will weather the quickest.

Answers to MATHBREAK

8 cm × 8 cm = 64 cm²

6 × 64 cm² = 384 cm²

MISCONCEPTION ALERT

Students may assume that some types of rock, such as granite, do not weather. Emphasize that all stone weathers, but different kinds of rock weather at different rates. The granite that is used on buildings and monuments is often polished. Polishing the surface slows the weathering process because less surface area is exposed.

Teaching Transparency 139 "Weathering and Surface Area"

Directed Reading Worksheet Section 2

internetconnect

SCiLINKS.
NSTA

TOPIC: Rates of Weathering
GO TO: www.scilinks.org
*sci*LINKS NUMBER: HSTE233

3) Close

Quiz

1. Different types of rock weather at different rates, true or false? (true)

2. Chemical weathering has no effect on the rate of mechanical weathering, true or false? (false)

3. What factors contribute to accelerated weathering rates at high elevations? (wind, precipitation, and gravity)

ALTERNATIVE ASSESSMENT

Investigate Your Area A cemetery is a great place to observe the effects of differential weathering because several different kinds of rock are used to make headstones, and most of the headstones are dated. Schedule a field trip to a cemetery, or encourage interested students to visit a cemetery on their own. Have them compare the dates and types of stone used to determine which kinds of stone are most susceptible to weathering.

Weathering and Climate

Imagine that two people have the same kind of bicycle. The frames of both bikes are made of steel. One person lives in a hot, dry desert in New Mexico, and the other lives on the warm, humid coastline of Florida, as shown in **Figure 15.** Both bicycles are outside all the time. Which bike do you think will have more problems with rust?

If you think the Florida bike will have more rust problems, you are right! Rust is iron oxide, and oxidation occurs more quickly in warm, humid climates. This is true for bikes, rocks, or anything else that is affected by chemical weathering.

Figure 15 *The climate in which this bike is located affects the rate at which it weathers, or rusts.*

Weathering and Elevation

When a newly formed mountain range extends up into the atmosphere, it tends to be weathered back down again. Weathering occurs in the same way on mountains as it does everywhere else, but rocks at high elevations are exposed to more wind, rain, and ice than rocks at lower elevations.

Gravity also takes its toll. The steepness of mountain slopes strengthens the effects of mechanical and chemical weathering. Rainwater quickly runs off the sides of mountains, carrying sediment with it. This continual removal of sediment exposes fresh rock surfaces to the effects of weathering. When rocks fall away from the sides of mountains, new surfaces are exposed to weathering. As you have learned, the greater the surface area is, the faster weathering occurs. If new mountain ranges didn't keep forming, eventually there would be no mountains at all!

---

internet connect

sciLINKS
NSTA

TOPIC: Rates of Weathering
GO TO: www.scilinks.org
*sci*LINKS NUMBER: HSTE233

SECTION REVIEW

1. How does surface area affect the rate of weathering?

2. How does climate affect the rate of weathering?

3. **Making Inferences** Does the rate at which a rock undergoes chemical weathering increase or stay the same when the rock becomes more mechanically weathered? Why?

38

▼ *Answers to Section Review*

1. Increasing the surface area of a rock increases the rate at which it weathers.

2. Answers will vary. In hot, dry climates, weathering happens more slowly than in hot, humid climates. This is because the presence of water increases the rate of weathering. Temperature extremes also accelerate weathering rates.

3. Mechanical weathering breaks down rock into smaller pieces, which increases its surface area. This exposes more surface area to the effects of chemical weathering.

From Bedrock to Soil

Terms to Learn

soil   humus
bedrock   topsoil
parent rock   leaching

What You'll Do

◆ Define *soil*.
◆ Explain the difference between residual and transported soils.
◆ Describe the three soil horizons.
◆ Describe how various climates affect soil.

What is soil? The answer depends on who you ask. A farmer may have a different answer than an engineer. To a scientist, **soil** is a loose mixture of small mineral fragments and organic material. The layer of rock beneath soil is called **bedrock.**

Sources of Soil

Not all soils are the same. In fact, soils differ from one another in many ways. Because soils are made from weathered rock fragments, the type of soil that forms depends on the type of rock that weathers. For example, the soil that forms from granite will be different from the soil that forms from limestone. The rock that is the source of soil is called **parent rock.**

Figure 16 shows a layer of soil over bedrock. In this case, the bedrock is the parent rock because the soil above it formed from the bedrock below. Soil that remains above the bedrock from which it formed is called *residual soil*. Notice the trees growing in this soil. Plants and other organisms, plus chemical weathering from water, help break down the parent rock into soil.

After soil forms, it can be blown or washed away from its parent rock. Once the soil is deposited, it is called *transported soil*. **Figure 17** shows one way that soil is transported from one place to another. The movement of glaciers is also responsible for deposits of transported soil.

Living Things Also Add to Soil In addition to bits of rock, soils also contain very small particles of decayed plant and animal material called **humus** (HYU muhs). In other words, humus is the organic part of the soil. Humus contains nutrients necessary for plant growth. In general, soil that contains as much as 20–30 percent humus is considered to be very healthy soil for growing plants.

Figure 16 *Residual soil is soil that rests on top of its parent rock.*

Figure 17 *Transported soil may be moved long distances from its parent rock by rivers such as this one.*

39

Homework

Ask students to find out why color is an important indication of soil composition. Have students make a poster showing the range of color possibilities for soils and what those colors indicate. Students should also take soil samples from different locations and write a paragraph about the origin of each sample, based on color, composition, particle size, texture, and pH.

internet**connect**

SC*LINKS*
NSTA

TOPIC: Soil and Climate
GO TO: www.scilinks.org
*sci*LINKS NUMBER: HSTE235

TOPIC: Soil Types
GO TO: www.scilinks.org
*sci*LINKS NUMBER: HSTE245

Focus

From Bedrock to Soil

In this section students learn how soil forms and investigate the links between organisms and soil formation. The section discusses how soil forms in layers of different composition, called horizons. Finally, students learn about the role climate plays in soil formation.

🔔 Bellringer

Ask students "Where does soil come from? Has there always been soil on Earth? Can soil become rock? What makes soil valuable to humans?" Have students write their ideas in their ScienceLog.

1) Motivate

GROUP ACTIVITY

Provide each group with magnifying lenses and samples of several types of local soil. Have students empty each sample onto a piece of white paper and inspect it closely by touching it and examining it with a magnifying lens. Have students record their observations about each sample's composition, color, particle size, texture, smell, and moisture content. Ask groups to hypothesize about how each soil type formed and about what type of plant life might grow in the soil. Sheltered English

Directed Reading Worksheet Section 3

MATERIALS

FOR EACH GROUP:
- soil samples (including potting soil)
- resealable plastic bags
- sterile spatula and tongs
- several pieces of bread (without preservatives)

Along with bacteria, fungi are the major decomposers of organic material in soil. Soil fungi excrete enzymes that break down organic matter into simpler substances that they can absorb. This process plays a crucial role in the circulation of nutrients at Earth's surface.

Have groups bring in several varieties of soil, such as clay, loam, and silt. To prevent contamination, emphasize that all materials should be handled with sterilized instruments. Have students put 20–30 drops of distilled water on each of the bread slices. Using the spatula, students should then sprinkle a small portion of each soil sample onto a piece of bread, one sample per piece of bread. Next, students should use the tongs to place each piece of bread in a separate, labeled plastic bag. One piece of bread with no soil should be used as a control. Place the bags in a dark box or drawer. After 5–7 days, students can use a stereomicroscope or a compound microscope to analyze the mold grown from each sample. Have students sketch what they observe and draw conclusions about the concentrations of organic materials and fungi in each sample.

Teaching Transparency 140
"Soil Horizons"

Soil Layers

As you've already learned, much of the material in residual soil comes from the bedrock that lies below it. Because of the way it forms, soil often ends up in a series of layers, with humus-rich soil on top, sediment below that, and bedrock on the bottom. Geologists call these layers *horizons*. The word *horizon* tells you that the layers are horizontal. **Figure 18** shows what these horizons can look like. You can see these layers in some road cuts.

The top layer of soil is often called the **topsoil**. Topsoil contains more humus than the layers below it, so it is rich in the nutrients plants need in order to be healthy. This is why good topsoil is necessary for farming. Topsoil is in limited supply because it can take hundreds and even thousands of years to form.

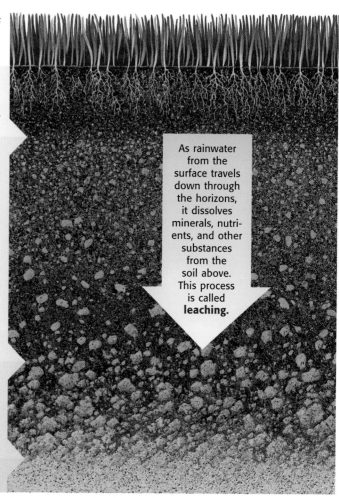

Figure 18 *This is what the layers of soil might look like if you dug a hole down to bedrock.*

Horizon A This layer consists of the topsoil. Topsoil contains more humus than any other soil horizon.

Horizon B This layer is often called the *subsoil.* This is where clays and dissolved substances from horizon A collect. This layer contains less humus.

Horizon C This layer consists of partially weathered bedrock, which is usually the parent rock of the soil above.

Bedrock

As rainwater from the surface travels down through the horizons, it dissolves minerals, nutrients, and other substances from the soil above. This process is called **leaching.**

40

**CONNECT TO
LIFE SCIENCE**

Students may be surprised to learn that soil is a thriving ecosystem. In 1 m^3 of soil, there may be 10 million roundworms and 50,000 small insects and mites. In a single gram of fertile soil, there may be 50,000 algae, 400,000 fungi, and 2.5 million bacteria. Have students construct a Berlese funnel to collect small organisms from soil.

IS THAT A FACT!

Where do soda cans and airplanes come from? The soil of course! The process of soil leaching produces concentrated bauxite deposits in a thin layer at Earth's surface. Bauxite is the ore that is refined to produce aluminum.

Soil and Climate

Soil types vary from place to place. As you know, this is partly due to differences in parent rocks. But it is also due to differences in climate. As you read on, you will see that climate can make big differences in the types of soil that develop around the world.

Tropical Climates Take a look at **Figure 19.** In tropical climates the air is very humid and the land receives a large amount of rain. You might think that a lot of rain always leads to good soil for growing plants. But remember that as water moves through the soil, it leaches material from the topsoil downward. Heavy rains cause this downward movement to occur quickly and constantly. The result is that tropical topsoil is very thin.

The vegetation growing in the topsoil keeps heavy rains from eroding it away. The vegetation, in turn, depends on the thin topsoil because the subsoil will not support lush plant growth. Agricultural and mining practices that disrupt this fragile balance can expose the topsoil to erosion. Once the topsoil is gone, the original plants will not return, and the production of topsoil will stop.

Figure 19 *The soil in tropical rain forests supports some of the lushest vegetation on Earth. However, tropical topsoil is extremely thin.*

Desert Climates While tropical climates get lots of rain, deserts get very little. Because of the lack of rain, deserts have very low rates of chemical weathering. When ground water trickles in from surrounding areas, some of it seeps upward. But as soon as the water gets close to the surface, it evaporates. This means that any materials that were dissolved in the water get left behind in the soil.

Often the chemicals left behind are various types of salts. The salts can sometimes get so concentrated that the soil becomes toxic, even to desert plants! This is one of the reasons for Death Valley's name. **Figure 20** shows the floor of Death Valley, in California.

Figure 20 *Very few plants can survive the harsh conditions of desert soils.*

CONNECT TO
ENVIRONMENTAL SCIENCE

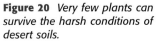

Writing Some windswept deserts have very little surface soil because wind has carried off most of the smaller particles. This leaves an exposed layer of pebbles and gravel too heavy to be moved by the wind. This layer, called desert pavement, may take hundreds of years to form, but once established, it protects the desert from further erosion. Desert pavement is easily destroyed by off-road vehicles. Ask students to write a persuasive essay arguing that such sensitive desert areas should or should not be off limits to vehicles.

BRAIN FOOD

In some desert areas, a special type of soil called cryptogamic soil is actually alive! This soil is composed of different species of mosses, lichens, fungi, and algae. Cryptogamic soil is sometimes known as "brown sugar soil" because it is dark brown and crusty. The spongy soil absorbs moisture readily, and when disturbed by freezing, it uplifts and cracks. The cracks are important to desert ecosystems because plant seeds get lodged in the cracks, and the moisture allows them to germinate. Cryptogamic soils are continually disrupted by cycles of freezing and thawing, but they can be severely damaged if they are walked on. Ask students to find out why walking on cryptogamic soil could damage it.

Multicultural CONNECTION

The Lacandon Maya of Mexico have developed sustainable farming methods that do not destroy the fragile soil of the tropical rain forest. On a small piece of land they grow both food crops and tree crops, a practice known as agroforestry. After a few years, they let the farmland recover by allowing it to become a forest again. The Lacandon Maya's approach to farming is recognized for its ecological soundness and has been replicated in many countries. This approach only works when a population has a lot of land area to spare so that some can lie fallow. Have students research sustainable farming techniques and create a model or poster to share with the class.

3 Extend

GOING FURTHER

Have student groups collect two soil samples from the same area, one from the surface and one from 16–20 cm down. Groups should fill two test tubes about one-quarter full with each soil sample and add water until the tubes are three-quarters full. Have students gently shake the covered tubes for several minutes. Place the test tubes in a rack and leave them overnight. Students should be able to observe the different compositions of layers that formed in the two test tubes. Soil components will settle according to weight; humus will be on top. The surface soil will probably contain noticeably more humus than the below-surface soil. Students might also notice a separation by grain size, with larger grains on the bottom.

4 Close

Quiz

1. The source of mineral fragments in soil is called the (parent rock).

2. The organic part of soil is called (humus).

3. What practices threaten the topsoil in tropical biomes? (agriculture and mining)

ALTERNATIVE ASSESSMENT

Writing Have students imagine that they are on a world trip during which they travel to every climate mentioned in the section. Tell them to write a series of postcards in which they describe what the soil is like in each climate. The picture on each card should be a magazine photograph that illustrates the soil in that climate.

Figure 21 *The rich soils in areas with a temperate climate support a vast farming industry.*

Figure 22 *Arctic soils, such as the soil along Denali Highway, in Alaska, cannot support lush growth.*

Temperate Climates Much of the continental United States has a temperate climate. An abundance of both mechanical and chemical weathering occurs in temperate climates. Temperate areas get enough rain to cause a high level of chemical weathering, but not so much that the nutrients are leached out. As a result, thick, fertile soils develop, as you can see in **Figure 21.**

Temperate soils are some of the most productive soils in the world. In fact, the midwestern part of the United States has earned the nickname "breadbasket" for the many crops the region's soil supports.

Arctic Climates You might not think that cold arctic climates are at all like desert climates. But many arctic areas have so little precipitation that they are actually cold deserts. As in the hot deserts, chemical weathering occurs very slowly, which means that soil formation also occurs slowly. This is why soil in arctic areas tends to be thin and is unable to support many plants, as shown in **Figure 22.**

internet connect

SCiLINKS
NSTA

TOPIC: Soil and Climate
GO TO: www.scilinks.org
*sci*LINKS NUMBER: HSTE235

SECTION REVIEW

1. What is the difference between residual and transported soils?

2. Which layer of soil is the most important for growing crops? Explain.

3. **Identifying Relationships** In which type of climate would leaching be more common—tropical or desert? Explain.

▼ Answers to Section Review

1. Residual soils form from the weathering of the parent rock beneath them; transported soils form in one place and are carried by wind or water to another place.

2. The topsoil (horizon A) is most important for growing crops because it contains the organic materials that crops need to grow.

3. Leaching would be more common in a tropical climate because more water passes through the soil.

Terms to Learn

soil conservation
erosion

What You'll Do

◆ Describe three important benefits that soil provides.
◆ Describe three methods of preventing soil erosion.

Soil Conservation

If we do not take care of our soils, we can ruin them or even lose them altogether. Many people assume that if you simply plow a field and bury some seeds, plants will grow. They also assume that if you grew a crop last year, you can grow it again next year. These ideas might seem reasonable at first, but farmers and others involved with agriculture know better. Soil is a resource that must be conserved. **Soil conservation** consists of the various methods by which humans take care of the soil. Let's take a look at why soil is so important and worth conserving.

The Importance of Soil

Consider some of the benefits of soil. Soil provides minerals and other nutrients for plants. If the soil loses these nutrients, then plants will not be able to grow. Take a look at the plants shown in **Figure 23**. The plants on the bottom look unhealthy because they are not getting enough nutrients. Even though there is enough soil to support their roots, the soil is not providing them with the food they need. The plants on the top are healthy because the soil they live in is rich in nutrients.

Poor agricultural practices often cause rich soils to lose their nutrients. It is important to have healthy soil in order to have healthy plants. All animals get their energy from plants, either directly or indirectly.

Housing Soil also provides a place for animals to live. Earthworms, grubs, spiders, moles, and prairie dogs all live in soil. If the soil disappears, so do the homes of these animals. These animals are also important to the soil and to plant growth because they help break down plant and animal matter to make humus.

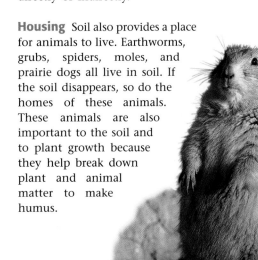

Figure 23 *Both photos above show the same crop. But the soil in the bottom photo is depleted of its nutrients.*

43

CONNECT TO
LIFE SCIENCE

Loam refers to a type of soil that, depending on the amount of humus, is best for plant growth. Because loam contains an ideal balance of different-size particles (sand, silt, and clay), the soil can retain the air and water essential for plant growth. Soils that are sandy have poor water-retention properties because the soil particles have too much space between them. Claylike soils lack air spaces and may be impervious to water. Have students research and report on some of the ways that plants have adapted to live in inhospitable soils.

Focus

Soil Conservation

This section discusses the fundamental importance of soil to all forms of life. Students learn that soil provides nutrients for plants to grow and stores ground water. They then explore some methods to prevent nutrient loss and erosion in topsoil.

⦿ Bellringer

Tell students that soil has been called "the bridge between life and the inanimate world." Then share Franklin D. Roosevelt's quote: "The nation that destroys its soil destroys itself." Have students write a ScienceLog entry that explores the meaning of both quotes.

1 Motivate

DISCUSSION

Students may think that all soils are merely dirt. Soils have different characteristics, depending on their composition. Engineers study soil types when planning roads and buildings. Different types of soils require different engineering considerations. For example, soils high in clay swell with rainfall and contract when they dry. The expanding and contracting can cause shifting and cracking in roadbeds and building foundations. If possible, arrange for a county extension agent, a soil conservationist, a geologist, or an engineer to speak with the class about the importance of understanding soil types.

Directed Reading Worksheet Section 4

ACTIVITY

Obtain three 4-in. flowerpots. Fill one with gravel, one with clay, and one with potting soil. Place them on a plank of wood over a sink. Have three students slowly pour 500 mL of water into each pot simultaneously. Ask students to note which pot retains the most water. Explain that the gravel started to leak first because the spaces between gravel are too large to hold the water. The potting soil has much smaller spaces, where water can be stored. The clay has very small spaces and will not allow water to pass through. Have students then explain in writing which material they would choose for the following activities:

- lining the foundation of a house so it drains quickly (gravel)
- sowing grass seed (potting soil)
- lining the bottom of an artificial pond so that it doesn't leak (clay)
  Sheltered English

Multicultural CONNECTION

By the early twentieth century, a century of cotton cultivation in the southern United States had so depleted the soil of nutrients that the area faced an agricultural crisis. An African-American scientist named George Washington Carver convinced farmers to plant peanuts and soybeans instead of cotton. Carver knew that these crops would restore a key nutrient—nitrogen—to the soil. The soil recovered, and Carver's work helped revitalize the agricultural economy of the South. Encourage students to learn more about the life and contributions of this remarkable scientist.

Figure 24 *When it rains, soil helps to store water that can later be used by plants and animals. When soil is removed or covered over, rainwater drains away.*

Storage Another benefit of soil is that it holds water, as shown in **Figure 24.** You might think of reservoirs, lakes, or even large tanks as places where water is stored. But soil is also extremely important in storing water. When water cannot sink into the ground, it quickly flows off somewhere else. Now that we have looked at the importance of soil, let's look at some ways we can maintain soil.

Figure 25 *Some of the topsoil that was once in this field has eroded, and the subsoil that remains is less able to support plant growth.*

Preventing Soil Erosion

Erosion is the process by which wind and water transport soil and sediment from one location to another. When soil is left unprotected, it is subject to erosion. **Figure 25** shows a field that has been stripped of part of its topsoil because no plants were growing in it. So while plants need soil to grow, plants are needed to keep topsoil from being eroded by wind and water. Soil conservation practices, like those discussed on the next page, help ensure that the soil is preserved for generations to come.

Homework

Indigenous Agriculture The ancient Maya of Central America used specialized agricultural techniques to maximize their corn crops. Because the Maya lived in a tropical area with karst topography, most of the soil naturally eroded into sinkholes or depressions. The Maya intensively planted in these locations. The soil above a sinkhole is ideal because it has the richest topsoil, and all surface water drains into the depressions. Have interested students find out more about the agricultural innovations of other indigenous cultures. Students may be surprised to learn that some of these techniques are still being used today.

Cover Crops Farmers can plant cover crops to prevent soil erosion. A *cover crop* is a crop that is planted between harvests to reduce soil erosion and to replace certain nutrients in the soil. Soybeans and clover are common cover crops.

Crop Rotation Fertile soil is soil that is rich in the nutrients that come from humus. If you grow the same crop year after year in the same field, certain nutrients become depleted. To slow this process down, crops can be changed from year to year. This practice, called *crop rotation,* is a common way to keep soils nutrient-rich.

Contour Plowing and Terracing How would you decide which direction to plow the rows in a field? If farmers plowed rows so that they ran up and down hills, what might happen during the first heavy rain? Hundreds of little river valleys would channel the rainwater down the hill, eroding the soil.

Take a look at the left-hand photo in **Figure 26.** Notice how the farmer has plowed across the slope of the hills instead of up and down the hills. This is called *contour plowing,* and it makes the rows act like a series of little dams instead of a series of little rivers. What if the hills are really steep? Farmers can use *terracing,* shown in the right-hand photo, to change one steep field into a series of smaller flat fields.

Figure 26 *Contour plowing and terracing are effective methods of preventing soil erosion.*

SECTION REVIEW

1. Describe three essential benefits that soil provides.

2. How does crop rotation benefit soil?

3. List three methods of soil conservation, and describe how each helps to prevent the loss of soil.

4. **Applying Concepts** Why do all animals, even meat eaters, depend on soil to survive?

internetconnect

SCI*LINKS*
NSTA

TOPIC: Soil Conservation
GO TO: www.scilinks.org
*sci*LINKS NUMBER: HSTE240

45

MATH and MORE

All soils contain varying amounts of water, air, minerals, and organic matter. A 200 cc soil sample ideal for plant growth may contain 40 cc of water, 50 cc of air, 70 cc of mineral fragments, and 40 cc of humus. Have students create a pie chart showing this composition by percentage.

Quiz

1. What is one way nutrients are depleted from soil? (when water passes through soil or when the same crops are repeatedly planted)

2. How do contour plowing and terracing help prevent soil erosion? (by interrupting water flow across the topsoil)

ALTERNATIVE ASSESSMENT

Declare "Soil Conservation Awareness Week." Have students create posters that alert your school to the importance of soil and highlight some ways to protect and conserve it.

 Critical Thinking Worksheet
"Buying the Farm"

 Reinforcement Worksheet
"Where the Tall Corn Grows"

 **internetconnect**

SCI*LINKS*
NSTA

TOPIC: Soil Conservation
GO TO: www.scilinks.org
*sci*LINKS NUMBER: HSTE240

▼ *Answers to Section Review*

1. Answers will vary. Soil provides nutrients for plants to grow and a home for many animals. Soil also stores water that both plants and animals need in order to live.

2. Crop rotation keeps the soil from becoming rapidly depleted of nutrients.

3. Answers will vary. The use of cover crops, contour plowing, and terracing are methods of soil conservation. Cover crops protect the soil from being eroded by wind and water. Contour plowing and terracing help to keep the soil from being eroded by running water.

4. Soil provides the nutrients that plants need to grow. All animals get their energy from plants, either directly or indirectly.

Rockin' Through Time
Teacher's Notes

Time Required

One 45-minute class period

Lab Ratings

EASY ———————————→ HARD

TEACHER PREP 🝮🝮
STUDENT SET-UP 🝮
CONCEPT LEVEL 🝮🝮
CLEAN UP 🝮

MATERIALS

The materials listed on the student page are adequate for groups of 4–5 students.

Safety Caution

Remind students to review all safety cautions and icons before beginning this lab activity. Be sure to use plastic bottles in this activity.

Larry Tackett
Andrew Jackson Middle School
Cross Lanes, West Virginia

Rockin' Through Time

Wind, water, and gravity constantly change rocks. As wind and water rush over the rocks, the rocks may be worn smooth. As rocks bump against one another, their shapes change. The form of mechanical weathering that happens as rocks collide and scrape together is called *abrasion*. In this activity, you will shake some pieces of limestone to model the effects of abrasion.

- poster board
- marker
- 24 pieces of limestone, all about the same size
- 3 L plastic, wide-mouth bottle with lid
- tap water
- graph paper or computer

Ask a Question

1 How does abrasion break down rocks? How can I use this information to identify rocks that have been abraded in nature?

Conduct an Experiment

2 Copy the chart on the next page onto a piece of poster board. Allow enough space to place rocks in each square.

3 Lay three of the limestone pieces on the poster board in the area marked "0 shakes." Be careful not to bump the poster board once you have added the rocks.

4 Place the remaining 21 rocks in the 3 L bottle. Then fill the bottle halfway with water.

5 Close the lid of the bottle securely. Shake the bottle vigorously 100 times.

6 Remove three rocks from the bottle, and place them on the poster board in the box labeled for the number of times the rocks have been shaken.

7 Repeat steps 5 and 6 six times until all of the rocks have been added to the board.

46

Make Observations

8 Describe the surface of the rocks that you placed in the area marked "0 shakes." Are the rocks smooth or rough?

9 How did the shapes of the rocks change as you performed this activity?

10 Using graph paper or a computer, construct a graph, table, or chart that describes how the shapes of the rocks changed as a result of the number of shakes they received.

Analyze the Results

11 Why did the rocks change?

12 How did the water change during the activity? Why did it change?

13 What would happen if you did this experiment using a much harder rock, such as gneiss?

14 How do the results of this experiment compare with what happens in a river?

Rocks	
0 shakes	100 shakes
200 shakes	300 shakes
400 shakes	500 shakes
600 shakes	700 shakes

Answers

8. The surfaces of the rocks are rough and jagged.

9. As the rocks were shaken more, they became much smoother.

11. The rocks became smoother because the edges broke off of them when they collided in the jar.

12. At the beginning of this activity, the water was clear. As the activity progressed, the water became increasingly dirty. This happened because particles broke away from the rocks and were suspended in the water.

13. If a harder rock were used, similar results would require longer and harder shaking.

14. As pebbles and small particles are carried along with the river's water, they bounce and grind against other rocks. Eventually, the rocks become smooth.

 Datasheets for LabBook

Chapter Highlights

Chapter Highlights

SECTION 1

Vocabulary
- **weathering** *(p. 30)*
- **mechanical weathering** *(p. 30)*
- **abrasion** *(p. 31)*
- **chemical weathering** *(p. 33)*
- **acid precipitation** *(p. 33)*
- **oxidation** *(p. 35)*

Section Notes
- Mechanical weathering is the breakdown of rock into smaller pieces by physical means.
- Ice wedging is a process by which water flows into cracks in rock and expands as it freezes, enlarging the cracks.
- The roots of plants can grow into cracks in rocks, and the roots can enlarge the cracks as they grow.
- Gravity, water, and wind are agents of abrasion in mechanical weathering.
- The activities of plants and animals can mechanically weather rock.
- Chemical weathering is the breakdown of rock into smaller pieces by chemical means.
- Water can dissolve some rocks and minerals.
- Sulfuric and nitric acids from pollution can cause chemical weathering.
- Natural acids found in air and water and produced by plants can cause chemical weathering.
- Oxidation can cause chemical weathering when oxygen combines with iron and other metallic elements.

Labs
Great Ice Escape *(p. 94)*

☑ Skills Check

Math Concepts

A CUBE'S TOTAL SURFACE AREA A cube has six sides—each is an identical square. To find the total surface area of a cube, first find the area of one of its sides. Then multiply the area of the square by 6 to find the total surface area of the cube. What is the total surface area of a cube that is 10 cm wide and 10 cm tall?

> Area of a square $= l \times w$
> Area of a cube $= 6 \, (l \times w)$
>
> $10 \text{ cm} \times 10 \text{ cm} = 100 \text{ cm}^2$
> $6 \times 100 \text{ cm}^2 = 600 \text{ cm}^2$

Visual Understanding

DIFFERENTIAL WEATHERING When a volcano becomes extinct, molten rock solidifies beneath the surface, forming harder, more weather-resistant rock than the sides of the volcano are made of. Shown in Figure 13, Devils Tower is a dramatic example of differential weathering at work.

Lab and Activity Highlights

Rockin' Through Time `PG 46`

Great Ice Escape `PG 94`

Datasheets for LabBook
(blackline masters for these labs)

SECTION 2

Vocabulary
differential weathering (p. 36)

Section Notes
- The rate at which weathering occurs depends partly on the composition of the rock being weathered.
- The greater the surface area of a rock is, the faster the rate of weathering.
- Different climates promote different rates of weathering.
- Weathering usually occurs at a faster rate at higher elevations.

SECTION 3

Vocabulary
soil (p. 39)
bedrock (p. 39)
parent rock (p. 39)
humus (p. 39)
topsoil (p. 40)
leaching (p. 40)

Section Notes
- Soil is made up of loose, weathered material that can include organic material called humus.
- Residual soils rest on top of their parent rock, and transported soils collect in areas far from their parent rock.
- Soil usually consists of horizons, layers that are different from one another.
- Soil types vary, depending on the climate in which they form.

SECTION 4

Vocabulary
soil conservation (p. 43)
erosion (p. 44)

Section Notes
- Soils are important because they provide nutrients for plants, homes for animals, and storage for water.
- Soils need to be protected from nutrient depletion and erosion through the use of soil conservation methods.

SECTION 2

differential weathering the process by which softer, less weather-resistant rocks wear away, leaving harder, more weather-resistant rocks

SECTION 3

soil a loose mixture of small mineral fragments and organic material

bedrock the layer of rock beneath soil

parent rock rock that is the source of soil

humus very small particles of decayed plant and animal material in soil

topsoil the top layer of soil that generally contains humus

leaching the process by which rainwater dissolves and carries away the minerals and nutrients in topsoil

SECTION 4

soil conservation the various methods by which humans take care of the soil

erosion the removal and transport of material by wind, water, or ice

 Blackline masters of these Chapter Highlights can be found in the **Study Guide.**

 internet**connect**

GO TO: go.hrw.com

Visit the **HRW** Web site for a variety of learning tools related to this chapter. Just type in the keyword:

KEYWORD: HSTWSF

sci LINKS SM
N S T A
GO TO: www.scilinks.org

Visit the **National Science Teachers Association** on-line Web site for Internet resources related to this chapter. Just type in the *sci*LINKS number for more information about the topic:

TOPIC: Weathering	*sci*LINKS NUMBER: HSTE230
TOPIC: Rates of Weathering	*sci*LINKS NUMBER: HSTE233
TOPIC: Soil and Climate	*sci*LINKS NUMBER: HSTE235
TOPIC: Soil Conservation	*sci*LINKS NUMBER: HSTE240
TOPIC: Soil Types	*sci*LINKS NUMBER: HSTE245

49

Lab and Activity Highlights

LabBank

 Whiz-Bang Demonstrations
When It Rains, It Fizzes

Calculator-Based Labs
A Soil Study

EcoLabs and Field Activities,
Whether It Weathers (or Not)

 Long-Term Projects & Research Ideas,
Precious Soil

Chapter Review
Answers

USING VOCABULARY

1. The chemical breakdown of rocks and minerals into new substances is called chemical weathering. Mechanical weathering is the physical breakdown of rocks and minerals; it does not result in new substances.
2. Oxidation is a chemical reaction that occurs when an element combines with oxygen. Rust, or iron oxide, is the result of one type of oxidation.
3. Residual soil is the soil that remains directly above the parent rock it came from. When the residual soil is carried by wind, water, or ice and deposited in another place, it is called transported soil.
4. Any rock layer directly beneath the soil is called bedrock. The rock that is the source of the soil is called the parent rock.
5. Contour plowing is a soil conservation technique in which a farmer plows across the slope of a hill rather than up and down it. Terracing is a technique in which the slope of a steep hill is divided into smaller flat fields that resemble stair steps.

UNDERSTANDING CONCEPTS

Multiple Choice
6. b	11. c
7. c	12. c
8. a	13. b
9. d	14. c
10. a	15. c

Short Answer
16. Mechanical weathering is the physical process of breaking down rock and minerals into smaller and smaller pieces. Chemical weathering is a chemical reaction that breaks down rock and minerals by changing them into different substances.
17. Caves usually form in limestone. This is because limestone reacts with the acid found in ground water.

Chapter Review

USING VOCABULARY

For each pair of terms, explain the difference in their meanings.

1. chemical weathering/mechanical weathering
2. oxidation/iron oxide
3. residual soil/transported soil
4. parent rock/bedrock
5. contour plowing/terracing

UNDERSTANDING CONCEPTS

Multiple Choice

6. Weathering by abrasion is usually caused by
   a. animals, plants, and wind.
   b. wind, water, and gravity.
   c. ice wedging, animals, and water.
   d. plants, gravity, and ice wedging.

7. Two acids found in acid precipitation are
   a. hydrochloric acid and sulfuric acid.
   b. nitric acid and hydrochloric acid.
   c. sulfuric acid and nitric acid.

8. Rust is produced by the oxidation of
   a. iron.   c. aluminum.
   b. tin.   d. manganese.

9. An acid normally involved in the formation of caves is
   a. nitric acid.
   b. hydrofluoric acid.
   c. hydrochloric acid.
   d. carbonic acid.

10. The soil horizon that contains humus is
    a. horizon A.   c. horizon C.
    b. horizon B.

11. The soil horizon that is made up of partially broken bedrock is
    a. horizon A.   c. horizon C.
    b. horizon B.

12. Tropical soils have the
    a. thickest horizon B.
    b. thickest horizon A.
    c. thinnest horizon A.
    d. thinnest horizon B.

13. The humus found in soils comes from
    a. parent rock.   c. bedrock.
    b. plants and   d. horizon B.
       animals.

14. Contour plowing means plowing
    a. up and down the slope of a hill.
    b. in steps along a hill.
    c. across the slope of a hill.
    d. in circles.

15. The main reason farmers use crop rotation is to slow down the process of
    a. soil removal by wind.
    b. soil removal by water.
    c. nutrient depletion.
    d. soil compaction.

Short Answer

16. Describe the two major types of weathering.

17. In what type of rock do caves usually form?

Concept Mapping Transparency 10

Blackline masters of this Chapter Review can be found in the **Study Guide.**

18. Why is Devils Tower higher than the surrounding area?

19. What can happen to soil when soil conservation is not practiced?

Concept Mapping

20. Use the following terms to create a concept map: weathering, chemical weathering, mechanical weathering, abrasion, ice wedging, oxidation, soil.

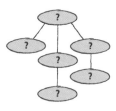

CRITICAL THINKING AND PROBLEM SOLVING

Write one or two sentences to answer the following questions:

21. Heat generally speeds up chemical reactions. But weathering, including chemical weathering, is usually slowest in hot, dry climates. Why is this?

22. How can too much rain deplete soil of its nutrients?

23. How does mechanical weathering speed up the effects of chemical weathering?

MATH IN SCIENCE

24. Imagine you are a geologist working in your natural laboratory—a mountainside. You are trying to find out the speed at which ice wedging occurs. You measure several cycles of freezing and thawing in a crack in a boulder. You discover that the crack gets deeper by about 1 mm per year. The boulder is 25 cm tall. Given this rate, how long will it take for ice wedging to split this boulder in half?

INTERPRETING GRAPHICS

The graph below shows how the density of water changes when temperature changes. The denser a substance is, the less volume it occupies. In other words, as most substances get colder, they contract and become more dense. But water is unlike most other substances—when it freezes, it expands and becomes less dense.

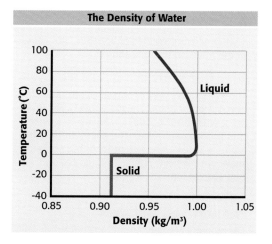

The Density of Water

25. Which will have the greater density, water at 40°C or water at −20°C?

26. How would the line in the graph look different if water behaved like most other liquids?

27. Which substance would be a more effective agent of mechanical weathering, water or some other liquid? Why?

Reading Check-up

Take a minute to review your answers to the Pre-Reading Questions found at the bottom of page 28. Have your answers changed? If necessary, revise your answers based on what you have learned since you began this chapter.

18. The rock that makes up Devils Tower is more resistant to weathering than the surrounding rock.

19. If soil conservation is not practiced, soil can be depleted of nutrients or eroded and transported away.

Concept Mapping

20. An answer to this exercise can be found at the front of this book.

CRITICAL THINKING AND PROBLEM SOLVING

21. Hot, dry climates generally have less precipitation than more temperate climates. Moisture enables chemical weathering to occur more quickly. The lack of moisture inhibits all processes of mechanical weathering except abrasion.

22. As water passes through soil, it carries the soil's nutrients with it. This process is called leaching.

23. Mechanical weathering increases the surface area of rock, thus exposing more of the rock to the effects of chemical weathering.

MATH IN SCIENCE

24. 1 cm = 10 mm
    25 cm = 250 mm
    250 mm/1 mm per year = 250 years

INTERPRETING GRAPHICS

25. water at 40°C

26. The line would slope downward from left to right.

27. Water is more effective than other liquids because it expands when it freezes.

ACROSS THE SCIENCES

Background

Students may be interested in researching the use of earthworms for composting organic wastes. Earthworms will consume yard and garden waste, food waste, and even cardboard and paper. Their castings can be harvested and used to provide organic fertilizer for yards or gardens. There are numerous Internet sites with detailed information on this topic.

EARTH SCIENCE • LIFE SCIENCE

Worms of the Earth

How much do you know about earthworms? Did you know they have no eyes and no ears? And did you know they can be as small as 1 mm or as long as 3 m? Earthworms and their relatives belong to the phylum Annelida. There are almost 12,000 different species of annelids, and some of them are pretty interesting!

Big, Old Worms

The *Rhinodrilus* earthworm of South America is about 2 m long and weighs about 1 kg. That's a big worm! But Australia is home to a worm that grows even bigger. The Gippsland earthworm is usually about 1 m long, but some of these worms have grown as long as 3 m. And some Gippsland earthworms have lived to be 10 years old!

Natural Soil Builders

Earthworms are very important to forming soil. As they dig through the soil searching for food, the tunnels they create expose rocks and minerals to the effects of weathering. Over time, this makes new soil. And as the worms tunnel, they mix the soil, allowing air and water and smaller organisms to move deeper into the soil.

Worms have huge appetites. They eat organic matter and other materials in the soil. One earthworm can eat and digest about half its body weight each day! This would be like someone who weighs 50 kg eating more than 25 kg of food each day! And eating all that food means that earthworms leave behind a lot of waste. Earthworm wastes, called castings, are very high in nutrients and make excellent natural fertilizer. Castings enrich the soil and enhance plant growth.

Making More Soil

Worms build and fertilize the soil, and plants grow. Plants then help make more soil. As roots grow and seek out water and nutrients, they help break larger rock fragments into smaller ones. Have you ever seen a plant growing in a crack in the sidewalk? As the plant grows, its roots spread into tiny cracks in the sidewalk. These roots apply pressure to the cracks, and over time, the cracks get bigger.

The same process occurs in rocks in the soil and on mountainsides. No matter where this process occurs, as the cracks expand, more water runs into them and more weathering takes place. Slowly, new soil is made. Sooner or later, maybe after hundreds of years, worms will be burrowing through what remains of that sidewalk or mountainside.

On Your Own

▶ Using the Internet and the library, do some research about earthworms and their relatives. Learn more about *Rhinodrilus* and the Gippsland earthworm. Or find out about leeches, a relative of earthworms. Some people even think earthworms would make tasty burgers—what do you think of that idea?

▼ *Notice the rings on this night crawler. The name* annelida *comes from the Latin word* **annellus**, *which means "ring."*

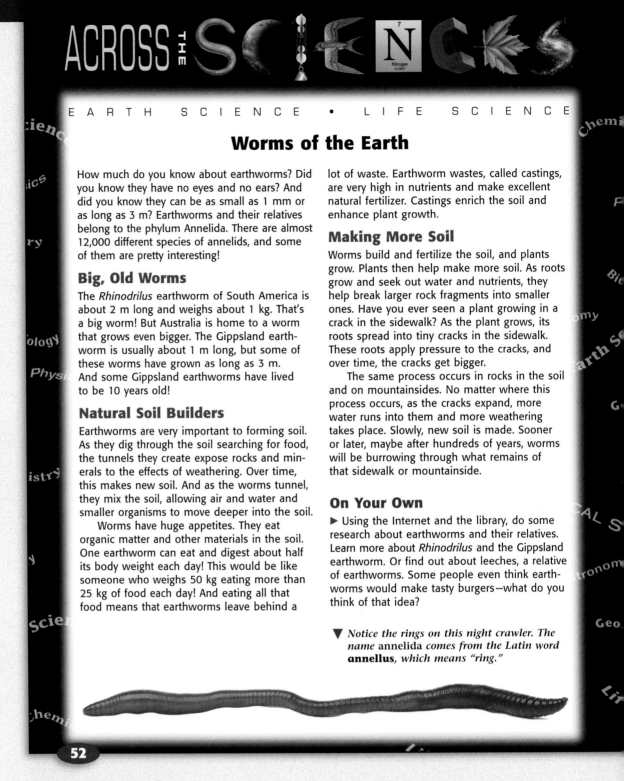

Answer to On Your Own

Gippsland earthworms are found in southeastern Australia. They live in clay soils found near rivers. People claim that a gurgling sound can be heard as the worms move down in their tunnels. *Rhinodrilus* is the largest land-dwelling segmented worm in the world. This South American earthworm is 2.5 cm thick, nearly 2 m long, and can weigh 1 kg. Parasitic leeches, which are related to earthworms, consume the blood of mammals and other vertebrates. Most land-dwelling parasitic leeches position themselves on low-growing plants and wait for suitable hosts to pass by. Some leeches sense body heat to find their prey. Other leeches climb trees to seek out mammals and birds. One South African species feeds on sea snails. These leeches not only kill the snails but liquefy them and suck up their flesh. Most leeches are compassionate parasites: leech saliva contains an antiseptic and a local anesthetic.

EYE ON THE ENVIRONMENT

Losing Ground

EYE ON THE
ENVIRONMENT

Losing Ground

▲ *In this example of sustainable farming, a new crop of soybeans grows up through the decaying remains of a corn crop.*

In the 1930s, massive dust storms in Oklahoma, Texas, Colorado, and New Mexico blew away the precious topsoil of many farms. Overplowing and a lack of rain caused this catastrophe—the Dust Bowl—that wreaked havoc on people's lives for 6 years. In the last 40 years, almost one-third of the world's topsoil has been lost to erosion. At the same time, the world's population has grown by 250,000 people per day. With more people to feed and less land to farm, some people are worried about having enough food to feed everyone.

Adding to the Problem

The topsoil on farmland is exposed to the full force of the weather. There are no trees to protect the loosely packed topsoil in a recently plowed field. Because of wind and water erosion, one hectare of farmland can lose more than 100,000 kg of soil in a year. Compare this loss to the mere 10–50 kg of soil lost in an average year by a forest densely packed with trees.

Tipping the Scales

In a healthy ecosystem, topsoil lost through erosion is replaced by other natural processes. These include the decomposition of organic matter by microorganisms and the breakdown of rocks by weathering. When a balance exists between the soil that is lost and the new soil that forms, the rate of topsoil loss is called sustainable. Currently, about 90 percent of the cropland in the United States is losing topsoil at a faster rate than is sustainable.

Sustainable Farming

The good news is that by changing their farming practices, many farmers have reduced the amount of soil lost from their fields. The critical step is to leave some plants growing in the ground. This protects the soil from the direct effects of wind and rain.

Many farmers have already switched to methods of sustainable farming. As the world's population continues to increase, more food will be needed. Because of this, preserving the topsoil that we have left will become more and more important.

On Your Own

▶ Find out what is meant by the term *desertification*. How does it relate to topsoil erosion?

Background

The Sahel region of northern Africa offers an excellent example of desertification. In the past, people in the drier part of the Sahel grazed animals, while those in the wetter part planted crops. The grazing animals were herded from place to place to find fresh grass and leaves. The cropland was planted only 4 or 5 years at a time, and then was allowed to lie fallow for several years. These methods allowed the land to support the people.

Today, population growth has led to overuse of the land. Several crops a year may be planted on the same land, and fallow periods are shortened or eliminated, causing the soil to lose its fertility. More and more animals are put out to graze. Trees and shrubs are cut for use as fuel or animal feed until few plants are left to hold the soil in place or to trap any rain that falls. The topsoil in the Sahel is carried away by wind and rain, and the land is becoming a desert.

Answer to On Your Own

The official definition for desertification (established at the 1992 Earth Summit) is "the land degradation in arid, semi-arid, and dry semi-humid areas resulting from various factors, including climate variations and human activities." One of the primary causes of desertification is overgrazing by animals. Today, fences prevent not only domestic animals but also wild animals from migrating to find food. Other causes of desertification are agricultural cultivation on marginal lands, the destruction of vegetation in semi-arid land, and poor irrigation practices. Desertification accelerates topsoil erosion because the land loses plant cover, exposing more soil to erosion.

Chapter Organizer

CHAPTER ORGANIZATION	TIME MINUTES	OBJECTIVES	LABS, INVESTIGATIONS, AND DEMONSTRATIONS
Chapter Opener pp. 54–55	45	National Standards: SAI 1, ES 2a	**Start-Up Activity,** Making Waves, p. 55
Section 1 Shoreline Erosion and Deposition	90	▶ Explain the connection between storms and wave erosion. ▶ Explain how waves break in shallow water. ▶ Describe how beaches form. ▶ Describe types of coastal landforms created by wave action. UCP 3, SAI 1, SPSP 2, 3, ES 1c	**Whiz-Bang Demonstrations,** Between a Rock and a Hard Place **Whiz-Bang Demonstrations,** Rising Mountains
Section 2 Wind Erosion and Deposition	90	▶ Explain why areas with fine materials are more vulnerable to wind erosion. ▶ Describe how wind moves sand and finer materials. ▶ Describe the effects of wind erosion. ▶ Describe the difference between dunes and loess. SAI 1, ST 2, SPSP 2, HNS 1–3, ES 1c, 2a; Labs UCP 2, 3, SAI 1	**QuickLab,** Making Desert Pavement, p. 63 **Making Models,** Dune Movement, p. 78 **Datasheets for LabBook,** Dune Movement
Section 3 Erosion and Deposition by Ice	90	▶ Summarize why glaciers are important agents of erosion and deposition. ▶ Explain how ice in a glacier flows. ▶ Describe some of the landforms eroded by glaciers. ▶ Describe some of the landforms deposited by glaciers. UCP 3, SAI 1, SPSP 3, ES 1c, 2a; Labs UCP 2, SAI 1, ES 1c	**Making Models,** Gliding Glaciers, p. 95 **Datasheets for LabBook,** Gliding Glaciers **Discovery Lab,** Creating a Kettle, p. 79 **Datasheets for LabBook,** Creating a Kettle
Section 4 Gravity's Effect on Erosion and Deposition	90	▶ Explain how slope is related to mass movement. ▶ State how gravity affects mass movement. ▶ Describe different types of mass movement. SAI 1, SPSP 3, 4, ES 1c	**QuickLab,** Angle of Repose, p. 74 **Long-Term Projects & Research Ideas,** Deep in the Mud

*See page **T23** for a complete correlation of this book with the*

NATIONAL SCIENCE EDUCATION STANDARDS.

TECHNOLOGY RESOURCES

 **Guided Reading Audio CD**
English or Spanish, Chapter 3

 One-Stop Planner CD-ROM with Test Generator

 Earth Science Videodisc
Reshaping the Crust: 44061–51180
Weathering and Erosion: 44062–46566
Glaciers and Erosion: 48922–51180

 CNN. Science, Technology & Society, Battling over the Oregon Inlet, Segment 17

Eye on the Environment, Shrinking Wetlands, Segment 8

CLASSROOM WORKSHEETS, TRANSPARENCIES, AND RESOURCES	SCIENCE INTEGRATION AND CONNECTIONS	REVIEW AND ASSESSMENT
Directed Reading Worksheet **Science Puzzlers, Twisters & Teasers**		
Math Skills for Science Worksheet, The Unit Factor and Dimensional Analysis **Directed Reading Worksheet,** Section 1 **Science Skills Worksheet,** Using Your Senses **Teaching Transparency 144,** Coastal Landforms Created by Wave Erosion: A **Teaching Transparency 145,** Coastal Landforms Created by Wave Erosion: B	**MathBreak,** Counting Waves, p. 57 **Connect to Physical Science,** p. 57 in ATE **Math and More,** p. 57 in ATE **Multicultural Connection,** p. 58 in ATE **Connect to Life Science,** p. 59 in ATE **Eye on the Environment:** Beach Today, Gone Tomorrow, p. 85	**Self-Check,** p. 57 **Section Review,** p. 61 **Quiz,** p. 61 in ATE **Alternative Assessment,** p. 61 in ATE
Transparency 146, Saltation **Directed Reading Worksheet,** Section 2 **Transparency 147,** Migration of Sand Dunes **Critical Thinking Worksheet,** A Future in Sand	**Connect to Life Science,** pp. 62, 63, 65 in ATE **Apply,** p. 64 **Multicultural Connection,** p. 64 in ATE **Cross-Disciplinary Focus,** p. 65 in ATE **Biology Connection,** p. 66 **Science, Technology, and Society:** Boulder Boogie, p. 84	**Self-Check,** p. 63 **Homework,** p. 65 in ATE **Section Review,** p. 66 **Quiz,** p. 66 in ATE **Alternative Assessment,** p. 66 in ATE
Directed Reading Worksheet, Section 3 **Math Skills for Science Worksheet,** Using Proportions and Cross-Multiplication **Transparency 148,** Landscape Features Carved by Alpine Glaciers **Reinforcement Worksheet,** An Alpine Vacation	**MathBreak,** Speed of a Glacier, p. 69 **Connect to Physical Science,** p. 69 in ATE **Cross-Disciplinary Focus,** pp. 70, 71 in ATE **Real-World Connection,** p. 72 in ATE **Cross-Disciplinary Focus,** p. 72 in ATE	**Self-Check,** p. 69 **Homework,** pp. 70, 71 in ATE **Section Review,** p. 73 **Quiz,** p. 73 in ATE **Alternative Assessment,** p. 73 in ATE
Transparency 220, The Law of Universal Gravitation **Directed Reading Worksheet,** Section 4	**Physics Connection,** p. 75 **Connect to Physical Science,** p. 75 in ATE **Real-World Connection,** p. 76 in ATE **Connect to Life Science,** pp. 76, 77 in ATE	**Homework,** p. 75 in ATE **Section Review,** p. 77 **Quiz,** p. 77 in ATE **Alternative Assessment,** p. 77 in ATE

END-OF-CHAPTER REVIEW AND ASSESSMENT

Chapter Review in Study Guide
Vocabulary and Notes in Study Guide
Chapter Tests with Performance-Based Assessment, Chapter 3 Test
Chapter Tests with Performance-Based Assessment, Performance-Based Assessment 3
Concept Mapping Transparency 12

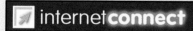

 Holt, Rinehart and Winston On-line Resources
go.hrw.com

For worksheets and other teaching aids related to this chapter, visit the HRW Web site and type in the keyword: **HSTICE**

 National Science Teachers Association
www.scilinks.org

Encourage students to use the *sci*LINKS numbers listed in the internet connect boxes to access information and resources on the **NSTA** Web site.

Chapter Resources & Worksheets

Visual Resources

TEACHING TRANSPARENCIES

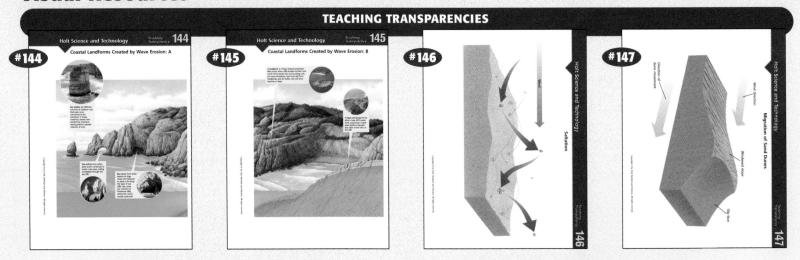

#144 — Holt Science and Technology — Teaching Transparency 144 — Coastal Landforms Created by Wave Erosion: A

#145 — Holt Science and Technology — Teaching Transparency 145 — Coastal Landforms Created by Wave Erosion: B

#146 — Holt Science and Technology — Saltation — Teaching Transparency 146

#147 — Holt Science and Technology — Migration of Sand Dunes — Teaching Transparency 147

TEACHING TRANSPARENCIES

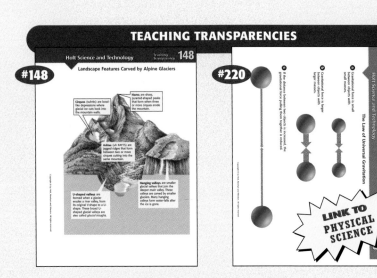

#148 — Holt Science and Technology — Teaching Transparency 148 — Landscape Features Carved by Alpine Glaciers

#220 — Holt Science and Technology — The Law of Universal Gravitation

LINK TO PHYSICAL SCIENCE

CONCEPT MAPPING TRANSPARENCY

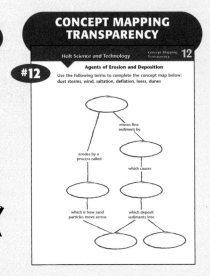

#12 — Holt Science and Technology — Concept Mapping Transparency 12 — Agents of Erosion and Deposition

Use the following terms to complete the concept map below: dust storms, wind, saltation, deflation, loess, dunes

Meeting Individual Needs

DIRECTED READING

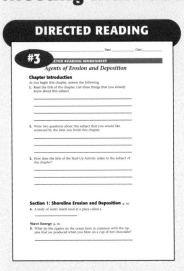

#3 — DIRECTED READING WORKSHEET — Agents of Erosion and Deposition

REINFORCEMENT & VOCABULARY REVIEW

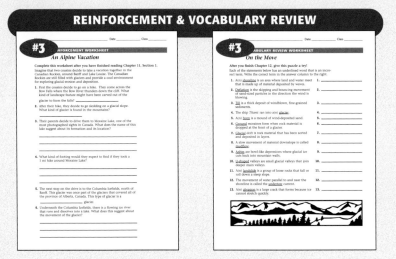

#3 — REINFORCEMENT WORKSHEET — An Alpine Vacation

#3 — VOCABULARY REVIEW WORKSHEET — On the Move

SCIENCE PUZZLERS, TWISTERS & TEASERS

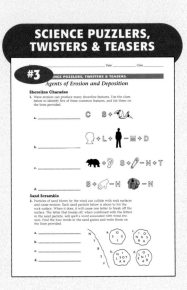

#3 — SCIENCE PUZZLERS, TWISTERS & TEASERS — Agents of Erosion and Deposition

Chapter 3 • Agents of Erosion and Deposition

Review & Assessment

STUDY GUIDE

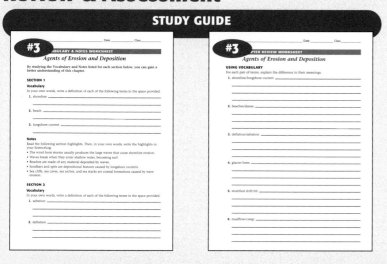

CHAPTER TESTS WITH PERFORMANCE-BASED ASSESSMENT

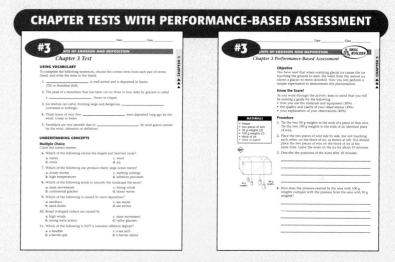

Lab Worksheets

WHIZ-BANG DEMONSTRATIONS

LONG-TERM PROJECTS & RESEARCH IDEAS

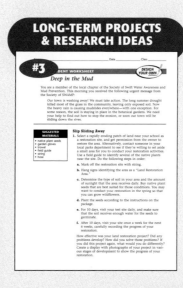

DATASHEETS FOR LABBOOK

Dune Movement

Gliding Glaciers

Creating a Kettle

Applications & Extensions

CRITICAL THINKING & PROBLEM SOLVING

SCIENCE TECHNOLOGY

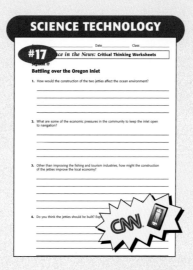

EYE ON THE ENVIRONMENT

Chapter Background

SECTION 1

Shoreline Erosion and Deposition

▶ Acrobatic Waves

To understand shoreline erosion, it's helpful to understand the forces acting in a breaking wave. Breaking waves can be thought of as somersaulting water. As waves move toward shallow coastal waters, the wavelengths shorten, crests crowd together, and wave height grows. When a wave becomes too top-heavy, it somersaults over itself, rushing onto the shore. As the water flows back into the ocean, it carries sand and sediment with it.

▶ The Origins of Cape Cod

At the end of the last glacial period—10,000 years ago— glaciers receding across North America helped form Cape Cod, Massachusetts. Cape Cod was initially mounds of outwash debris left behind by the glaciers. These mounds were then surrounded by the rising sea. Over time, currents eroded land and filled in depressions between the islands. Sandbars connected the islands to each other and the mainland. Since that time, Cape Cod has lost 3.2 km of coastline to ocean erosion.

IS THAT A FACT!

- ◆ Scientists predict that if the erosion continues at the current rate, Cape Cod will be completely reclaimed by the ocean in 4,000–5,000 years.

SECTION 2

Wind Erosion and Deposition

▶ The Dust Bowl

The Dust Bowl was a section of the Great Plains of the United States that extended from southeastern Colorado and southwestern Kansas to the panhandles of Texas and Oklahoma and to northeastern New Mexico. In the early 1930s, following years of overcultivation in the 1920s, the region suffered a severe drought. Exposed topsoil was carried away by strong spring winds. Windblown soil sometimes blocked out the sun, and the dirt piled up in drifts like snow. Occasionally, huge dust storms blew across the country and reached the East Coast. The wind erosion was gradually halted when the federal government planted windbreaks and large areas of grasslands were restored. The area had mostly recovered by the early 1940s.

▶ Lost Cities of the Takla Makan Desert

The Takla Makan Desert in China's arid northwest is so inhospitable that its name in the local language means "Go in, and you don't come out." The desert is covered with treacherous dunes of fine, dry sand. Buried under those dunes are the remains of cities that prospered along the ancient Silk Road. The Silk Road was a trade route that connected China to civilizations in the West. NASA's Spaceborne Imaging Radar (SIR-C), which flew on space shuttles twice in 1994, is being used to examine the desert. The radar-imaging technology has already helped archaeologists locate some cities and promises to help them find other ruins.

SECTION 3

Erosion and Deposition by Ice

▶ Glaciers and Drinking Water

About 10 percent of Earth's surface is covered with glaciers. The water frozen in glaciers makes up almost 75 percent of the world's freshwater supply. Arapaho Glacier, a small glacier in Colorado, provides water to more than 75,000 people living in the city of Boulder. Many countries have explored the possibility of obtaining drinking water from glaciers, even by towing icebergs to a nearby harbor!

IS THAT A FACT!

- ▰ Glaciers flow at different rates. Most glaciers flow at a rate of 1 m per day or less, but some flow much faster. In 1936, the Black Rapids Glacier, in Alaska, was measured flowing at a rate of 30 m per day.

- ▰ If all of Earth's glaciers simultaneously melted, sea level would rise more than 65 m, submerging coastal cities all over the world.

▶ Battles on Siachen Glacier

At 70 km, the Siachen Glacier is one of the world's longest. It's in the Karakoram Range, on the India–Pakistan border. It is also the site of the world's highest battles. Indian and Pakistani soldiers have fought over the disputed territory of Kashmir on peaks as high as 6,400 m (21,000 ft).

▶ The Great Lakes

The Great Lakes were formed by the movement of ice sheets during the Pleistocene epoch. These glaciers advanced over the land, gouging out a series of deep basins. As the glaciers melted, the basins filled with meltwater, and the five Great Lakes were formed.

SECTION 4

Gravity's Effect on Erosion and Deposition

▶ Scree

Mountain stones and boulders loosened by weathering and carried downward by gravity may be deposited in long, loose heaps called scree at the base of a mountain.

IS THAT A FACT!

- ▰ When an earthquake measuring 5 on the Richter scale occurred near Mount St. Helens on May 18, 1980, it triggered a landslide of more than 2 km³ of rock and ice. Immediately afterward, an eruption began, and an explosion of steam and volcanic gases produced a lahar that raced down the mountain at speeds of up to 250 km/h.

For background information about teaching strategies and issues, refer to the *Professional Reference for Teachers.*

CHAPTER 3

Agents of Erosion and Deposition

Pre-Reading Questions

Students may not know the answers to these questions before reading the chapter, so accept any reasonable response.

Suggested Answers

1. Answers may vary. Waves and wind both erode and deposit sediment. Students may also note that wind causes waves.

2. Answers may vary.

CHAPTER 3

Agents of Erosion and Deposition

Sections

Pre-Reading Questions

1. What do waves and wind have in common?

2. How do waves, wind, and ice erode and deposit rock materials?

54

THE CRASHING SURF

On February 8, 1998, unusually large waves crashed against the cliffs along Broad Beach Road in Malibu, California. Eventually, the ocean-eroded cliffs buckled, causing a landslide. One house collapsed into the ocean, while two more dangled on the edge of the cliff. What made these waves stronger than usual? How can water cause this much damage? In this chapter, you will study how waves shape beaches and coastlines. You will also learn how wind, moving ice, and gravity sculpt the surface of our planet.

---

internetconnect

HRW On-line Resources

go.hrw.com
For worksheets and other teaching aids, visit the HRW Web site and type in the keyword: **HSTICE**

SCILINKS
NSTA

www.scilinks.com
Use the *sci*LINKS numbers at the end of each chapter for additional resources on the **NSTA** Web site.

Smithsonian Institution

www.si.edu/hrw
Visit the Smithsonian Institution Web site for related on-line resources.

CNNfyi.com

www.cnnfyi.com
Visit the CNN Web site for current events coverage and classroom resources.

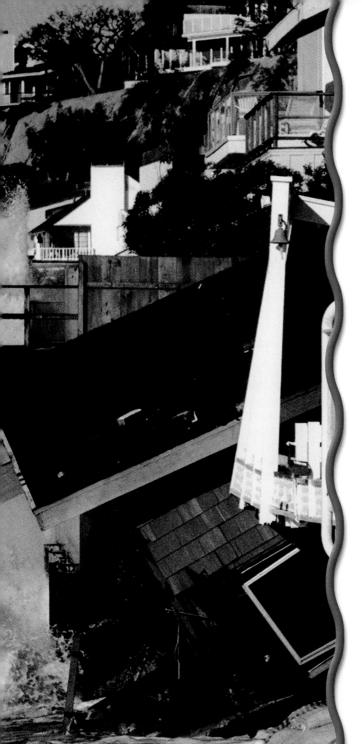

MAKING WAVES

Did you know that beaches and shorelines are shaped by crashing waves? See for yourself by creating some waves of your own.

Procedure

1. Make a beach by adding **sand** to one end of a **washtub.**

2. Fill the washtub with **water** to a depth of 5 cm.

3. In your ScienceLog, sketch the beach profile (side view), and label it "A."

4. Place a **block** at the end of the washtub opposite the beach. Move the block up and down very slowly to create small waves for 2 minutes. Sketch the new beach profile and label it "B."

5. Again place a block at the end of the washtub opposite the beach. Move the block up and down more rapidly to create large waves for 2 minutes. Sketch the new beach profile and label it "C."

Analysis

6. Compare the three beach profiles. What is happening to the beach?

7. How do small waves and large waves erode the beach differently?

8. What other factors might contribute to beach erosion?

55

START-UP Activity

MAKING WAVES

MATERIALS

FOR EACH GROUP:
- sand
- washtub
- tap water
- wooden or plastic block

Answers to START-UP Activity

6. The beach is slowly receding, or eroding.

7. Small waves erode less shoreline than large waves. Therefore, large waves have a greater impact on the shoreline.

8. Accept all reasonable responses. Sample answers: Lack of vegetation increases beach erosion. Wind and storms contribute to beach erosion.

Focus

Shoreline Erosion and Deposition

This section explores how wave action sculpts and builds shorelines. Students first focus on the formation of beaches and offshore landforms by deposition and then learn how waves erode the shoreline. Students explore the formation of sea cliffs, sea stacks, sea arches, and sea caves, and learn how waves work to erode cliffs, creating headlands and wave-cut terraces.

Bellringer

Writing Ask students to think about where sand comes from. Have them write a short poem in their ScienceLog about how ocean waves create sand from rock.

1) Motivate

ACTIVITY

Explain to students that shorelines are dynamic, changing environments because ocean waves and currents continually erode and redeposit sand. Have each student draw a "filmstrip" illustrating the changes that could occur in the history of a beach.

Encourage students to illustrate the processes that cause these changes and to write a caption that explains each frame and the time interval that elapses between scenes. Students can present their filmstrips to the class. **Sheltered English**

Terms to Learn

shoreline
beach
longshore current

What You'll Do

- ◆ Explain the connection between storms and wave erosion.
- ◆ Explain how waves break in shallow water.
- ◆ Describe how beaches form.
- ◆ Describe types of coastal landforms created by wave action.

Figure 1 *Waves produced by storms on the other side of the Pacific Ocean propel this surfer toward a California shore.*

56

MISCONCEPTION ALERT

It is a popular misconception that a wave is a moving wall of water. Water actually moves up and down rather than forward as wave energy travels through it. When waves break, however, as shown in **Figure 1,** they do carry water with them.

Shoreline Erosion and Deposition

What images pop into your head when you hear the word *beach*? You probably picture sand, blue ocean as far as the eye can see, balmy breezes, and waves. In this section you will learn how all those things relate to erosion and deposition along the shoreline. A **shoreline** is where land and a body of water meet. *Erosion,* as you may recall, is the breakdown and movement of materials. *Deposition* takes place when these materials are dropped. Waves can be powerful agents of erosion and deposition, as you will soon learn.

Wave Energy

Have you ever noticed the tiny ripples created by your breath when you blow on a cup of hot chocolate to cool it? Similarly, the wind moves over the ocean surface, producing ripples called *waves*. The size of a wave depends on how hard the wind is blowing and the length of time the wind blows. The harder and longer the wind blows, the bigger the wave is. Try it the next time you drink cocoa.

The wind that comes from severe winter storms and summer hurricanes generally produces the large waves that cause shoreline erosion. Waves may travel hundreds or even thousands of kilometers from a storm before reaching the shoreline. Some of the largest waves to reach the California coast are produced by storms as far away as Alaska and Australia. Thus, the California surfer in **Figure 1** can ride a wave produced by a storm on the other side of the Pacific Ocean.

Wave Trains On your imaginary visit to the beach, do you remember seeing just one wave? Of course not; waves don't move alone. They travel in groups called *wave trains*, as shown in **Figure 2.** As wave trains move away from their source, they travel through the ocean water without interruption. When they reach shallow water, they change form and begin to break. The ocean floor crowds the lower part of the wave, shortening the wave length and increasing the wave height. This results in taller, more closely spaced waves.

When the top of the wave becomes so tall that it cannot support itself, it begins to curl and break. These breaking waves are known as *surf.* Now you know how surfers got their name. The *wave period* is the time interval between breaking waves. Wave periods are usually 10 to 20 seconds long.

Figure 2 *Because waves travel in wave trains, they break at regular intervals.*

The Pounding Surf A tremendous amount of energy is released when waves break, as shown in **Figure 3.** A crashing wave can break solid rock or throw broken rocks back against the shore. The rushing water in breaking waves can easily wash into cracks in rock, helping to break off large boulders or fine grains of sand. The loose sand picked up by the waves polishes and wears down coastal rocks. Waves can also move sand and small rocks and deposit them in other locations, forming beaches.

Figure 3 *Breaking waves crash against the rocky shore, releasing their energy.*

MATH BREAK

Counting Waves

How many waves do you think reach a shoreline in a day if the wave period is 10 seconds?
(Hint: Calculate how many waves occur in a minute, in an hour, and in a day.)

> ### ✔ Self-Check
>
> Would a large wave or a small wave have more erosive energy? Why? *(See page 120 to check your answer.)*

57

Answer to Self-Check

A large wave has more erosive energy than a small wave because a large wave releases more energy when it breaks.

> ### CONNECT TO
> ### PHYSICAL SCIENCE

The waves in lakes and oceans are a form of energy traveling through a medium—water. Other energy waves, such as sound waves, also require a medium through which to travel. Some waves, however, such as light and radio waves, can travel through a vacuum.

MATH and MORE

Have students calculate how many waves would reach a shoreline in 24 hours with periods of 15 and 20 seconds.

(With a 15-second period, 5,760 waves would reach the shore. With a 20-second period, 4,320 waves would reach the shore.)

 Math Skills Worksheet "The Unit Factor and Dimensional Analysis"

Answer to MATHBREAK

In 1 minute, six waves occur. In 1 hour, 360 waves occur. In 1 day, 8,640 waves occur.

$$60 \div 10 = 6 \text{ waves}$$
$$6 \times 60 = 360 \text{ waves}$$
$$360 \times 24 = 8,640 \text{ waves}$$

 Directed Reading Worksheet Section 1

USING THE FIGURE

Draw students' attention to **Figure 5.** Point out that when water strikes the shoreline at an angle and then retreats in a direction perpendicular to the shore, material is moved along the beach in a zigzag pattern. Inform students that this is known as *beach* or *longshore drift*.

Encourage students to use the Internet or reference texts in their school library to learn more about longshore drift. Ask students to prepare labeled diagrams in their ScienceLog illustrating the phenomenon.
Sheltered English

Multicultural CONNECTION

The Polynesians are considered some of the greatest navigators of the ancient world. Polynesians visited and inhabited more than 10,000 islands throughout the South Pacific. They navigated not by using maps but by carefully observing stars, winds, and waves. On cloudy nights, they listened to the way the waves rocked and slapped against their dugout canoes. The Polynesians understood how wave patterns could indicate the direction of land or the presence of dangerous reefs or sandbars. Encourage students to discover more about Polynesian cultures in the past and present.

Science Skills Worksheet
"Using Your Senses"

Figure 4 *Beaches are made of different types of material deposited by waves.*

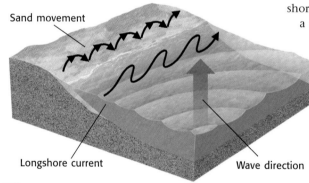

Sand movement

Longshore current

Wave direction

Wave Deposits

Waves carry an assortment of materials, including sand, rock fragments, and shells. Often this material is deposited on the shore. But as you will learn, this is not always the case.

Beaches You would probably recognize a beach if you saw one. But technically, a **beach** is any area of the shoreline made up of material deposited by waves. Some beach material arrives on the shoreline by way of rivers. Other beach material is eroded from areas located near the shoreline.

Not all beaches are the same. Compare the beaches shown in **Figure 4.** Notice that the colors and textures vary. This is because the type of material found on a beach depends on its source. Light-colored sand is the most common beach material. Much of this sand comes from the quartz in continental rock. But not all beaches are made of light-colored sand. For instance, on many tropical islands, beaches are made of fine white coral material, and some Florida beaches are made of tiny pieces of broken seashells. In Hawaii, there are black sand beaches made of eroded volcanic lava. In areas where stormy seas are common, beaches are made of pebbles and larger rocks.

Wave Angle Makes a Difference The movement of sand along a beach depends on the angle at which the waves strike the shore. Most waves approach the beach at a slight angle and retreat in a direction more perpendicular to the shore. This moves the sand in a zigzag pattern along the beach, as you can see in **Figure 5.**

Figure 5 *When waves strike the shoreline at an angle, sand migrates along the beach in a zigzag path.*

SCIENCE HUMOR

Q: Why does the beach think the ocean is friendly?

A: because it waves all the time

Offshore Deposits Waves moving at an angle to the shoreline push water along the shore, creating longshore currents. A **longshore current** is a movement of water near and parallel to the shoreline. Sometimes waves erode material from the shoreline, and a longshore current transports and deposits it offshore, creating landforms in open water. Some of these landforms are shown in **Figure 6.**

Figure 6 Sandbars and barrier spits are types of offshore deposits.

A **sandbar** is an underwater or exposed ridge of sand, gravel, or shell material.

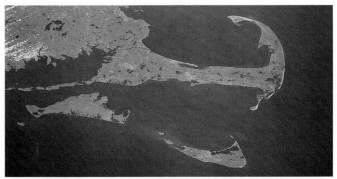

A **barrier spit,** like Cape Cod, Massachusetts, occurs when an exposed sandbar is connected to the shoreline.

Wave Erosion

Wave erosion produces a variety of features along a shoreline. *Sea cliffs,* like the ones in **Figure 7,** are formed when waves erode and undercut rock, producing steep slopes. Waves strike the base of the cliff, wearing away the soil and rock and making the cliff steeper. The rate at which the sea cliffs erode depends on the hardness of the rock and the energy delivered by the waves. Sea cliffs made of hard rock, such as granite, erode very slowly. Other sea cliffs, such as those made of soft sedimentary rock, erode rapidly, especially during storms.

Figure 7 *Ocean-view homes built on sedimentary rock are often threatened as cliffs erode.*

59

CONNECT TO LIFE SCIENCE

Beaches and intertidal zones can be a challenging place for organisms to live. Beaches offer little protection from predators, and the intertidal zone is periodically pounded by waves and exposed to the sun. Most of the organisms that live in these areas have special adaptations for survival. Have students research how different organisms are adapted for living in these environments. Suggest that students use their research to create a diorama that illustrates an intertidal zone.

BRAIN FOOD

Despite the numerous changes wrought by the sea, people still live and vacation as close to the water as possible. Inevitably, property is damaged. Government loan subsidies to these property owners cost taxpayers millions of dollars a year. Encourage students to consider the costs and benefits of erosion prevention. What solutions to the problem would they propose? How would they finance their plans? Allow time for them to share their ideas with the class.

IS THAT A FACT!

During storms and at high tide, waves deposit sand at the back of a sloping beach, forming a feature called a *berm.* Berms mark the highest point that waves reached during the last storm or high tide. Although berms may obstruct beachfront views, leveling them is not a good idea: they prevent the erosion of inland soil.

3) Extend

GROUP ACTIVITY

Coastal Features Board Game
To reinforce section concepts, divide the class into small groups, and challenge each to create a board game. Tell students that the object of the game is for players to visit as many coastal landforms as they can. Provide each group with poster board, plain index cards, and markers. Direct them to create a game board that leads players along a coastline, encountering the features they have learned about. Have students use the index cards to write questions and clues to direct players' movements along the "coast." For example, they might write, "If you can describe how the sea arch formed, you may move ahead to the sea stack. If not, you lose a turn." Have groups create written game rules, exchange games, and play their games.   **Sheltered English**

GOING FURTHER

Have students research and model different methods that have been used to control shoreline erosion. Encourage students to discover which methods have been effective and which have failed. Students can also develop their own plans for minimizing erosion and present them to the class.

Teaching Transparency 144
"Coastal Landforms Created by Wave Erosion: A"

internet connect

SCiLINKS NSTA

TOPIC: Wave Erosion
GO TO: www.scilinks.org
*sci*LINKS NUMBER: HSTE280

Shaping a Shoreline   Much of the erosion responsible for landforms you might see along the shoreline takes place during storms. Large waves generated by storms release far more energy on the shoreline than do normal waves. This energy is so powerful that it is capable of removing huge chunks of rock. The following illustrations show some of the major landscape features that result from wave erosion.

Coastal Landforms Created by Wave Erosion

Sea stacks are offshore columns of resistant rock that were once connected to the mainland. In these instances, waves have eroded the mainland, leaving behind isolated columns of rock.

Sea arches form when wave action continues to erode a sea cave, cutting completely through the rock.

Sea caves form when waves cut large holes into fractured or weak rock along the base of sea cliffs. Sea caves are common in limestone cliffs, where the rock is usually quite soft.

60

WEIRD SCIENCE

Sometimes, even when the weather is clear and calm, huge waves, called *rogue waves*, unexpectedly appear. These waves are responsible for damaging or sinking several ships a year. Rogue waves are a poorly understood phenomenon of the high seas. One reason so little is known about them is that their random nature makes them hard to study.

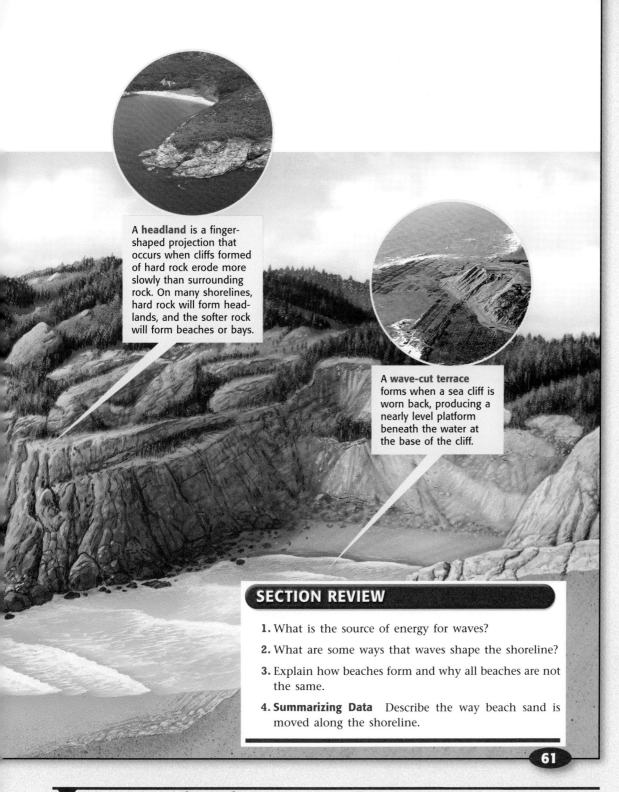

A **headland** is a finger-shaped projection that occurs when cliffs formed of hard rock erode more slowly than surrounding rock. On many shorelines, hard rock will form headlands, and the softer rock will form beaches or bays.

A **wave-cut terrace** forms when a sea cliff is worn back, producing a nearly level platform beneath the water at the base of the cliff.

SECTION REVIEW

1. What is the source of energy for waves?

2. What are some ways that waves shape the shoreline?

3. Explain how beaches form and why all beaches are not the same.

4. **Summarizing Data**   Describe the way beach sand is moved along the shoreline.

61

Quiz

1. What is a wave period? (It is the time it takes for two waves to pass a fixed point.)

2. What determines the way sand moves on a beach? (the direction in which waves strike the shore)

3. Describe how sea stacks, sea caves, and headlands are formed. (Sea stacks are columns of resistant rock left behind when a headland erodes. Sea caves form when waves erode large holes in fractured or weak rock at the base of sea cliffs. A headland forms when cliffs made of hard rock erode more slowly than the surrounding rock; this results in a finger-shaped projection.)

ALTERNATIVE ASSESSMENT

   Have students work independently to make a model of several land features created by waves. Ask them to present their models to the class and explain how the water strikes the shore to create the landforms. Have them brainstorm about what organisms, if any, would live on the landforms they modeled.

   Teaching Transparency 145 "Coastal Landforms Created by Wave Erosion: B"

▼ **Answers to Section Review**

1. Wind is the source of wave energy.

2. Sample answer: Waves erode rock, creating headlands and undercutting sea cliffs, sea caves, stacks, and sea arches. Waves also deposit material, forming landforms such as beaches, spits, and sandbars.

3. Beaches are usually made of geologic material eroded from areas of higher elevation. Rivers deposit this material in oceans. Waves and currents then redeposit the material along the shoreline to form beaches. All beaches are not the same because source materials vary.

4. Beach sand is moved by waves and by longshore currents. If the waves arrive at the shore at an angle, and retreat perpendicular to the shoreline, sand is deposited in a zigzag pattern. Longshore currents move sand in the direction they flow, parallel to the shore.

Focus

Wind Erosion and Deposition

In this section, students learn about the effects of wind erosion. They will explore the three major processes of wind erosion: saltation, deflation, and abrasion. Section 2 explains how the wind deposits materials, such as sand and loess. Students also learn about the migration of sand dunes.

🔔 Bellringer

Ask students to answer the following question in their ScienceLog:

What causes wind? (Students should understand that wind is caused by energy from the sun. The sun heats the Earth unevenly; the unequal energy distribution causes pressure differences which, in turn, cause air to move.)

1 Motivate

DISCUSSION

Wind Erosion Engage students in a discussion about the ways wind shapes the Earth's surface. Encourage them to compare the wind with waves. (They should recognize that both change the Earth's shape by erosion and deposition.)

Ask students to think of examples of the wind's effects on landscapes they have observed. (Answers might include the formation of sand dunes, wind-weathered surfaces, and so on.)

Point out that this section will explore the ways in which wind acts to erode and deposit materials on Earth.

Terms to Learn

saltation   dune
deflation   loess
abrasion

What You'll Do

◆ Explain why areas with fine materials are more vulnerable to wind erosion.
◆ Describe how wind moves sand and finer materials.
◆ Describe the effects of wind erosion.
◆ Describe the difference between dunes and loess deposits.

Have you ever tried to track a moving rock? Sounds silly, but in California some rocks keep sneaking around. To find out more, turn to page 84.

Wind Erosion and Deposition

Most of us at one time or another have been frustrated by a gusty wind that blew an important stack of papers all over the place. Remember how fast and far the papers traveled, and how it took forever to pick them up because every time you caught up with them they were on the move again? If you are familiar with this scene, then you already know how wind erosion works. Certain locations are more vulnerable to wind erosion than others. Areas with fine, loose rock material that have little protective plant cover can be significantly affected by the wind. Plant roots anchor sand and soil in place, reducing the amount of wind erosion. The landscapes most commonly shaped by wind processes are deserts and coastlines.

Process of Wind Erosion

Wind moves material in different ways. In areas where strong winds occur, material is moved by saltation. **Saltation** is the movement of sand-sized particles by a skipping and bouncing action in the direction the wind is blowing. As you can see in **Figure 8,** the wind causes the particles to bounce. When bouncing sand particles knock into one another, some particles bounce up in the air and fall forward, striking other sand particles. The impact may in turn cause these particles to roll forward or bounce up in the air.

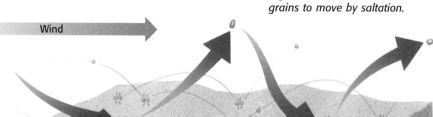

Figure 8 *The wind causes sand grains to move by saltation.*

Wind

Point out to students that many desert animals have special adaptations to protect themselves from windblown sand. For example, some lizards have transparent eyelids that shield their eyes from blowing sand while still allowing them to see.

Encourage students to use their school library to learn more about animal adaptations that protect against blowing sand. Have them prepare brief oral reports to share their findings with the class.

Deflation The lifting and removal of fine sediment by wind is called **deflation.** During deflation, wind removes the top layer of fine sediment or soil, leaving behind rock fragments that are too heavy to be lifted by the wind. This hard, rocky surface, consisting of pebbles and small broken rocks, is known as *desert pavement.* An example is shown in **Figure 9.**

Figure 9 *Desert pavement, like that found in the Painted Desert, in Arizona, forms when wind removes all the fine materials.*

Have you ever blown on a layer of dust while cleaning off a dresser? If you have, you might have noticed that in addition to your face getting dirty, a little scooped-out depression formed in the dust. Similarly, where there is little vegetation, wind may scoop out depressions in the sand. These depressions, like the one shown in **Figure 10,** are known as *deflation hollows.*

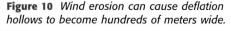

Figure 10 *Wind erosion can cause deflation hollows to become hundreds of meters wide.*

63

Quick Lab

Making Desert Pavement

1. Spread a mixture of **dust, sand,** and **gravel** on an outdoor **table.**

2. Place an **electric fan** at one end of the table.

3. Put on **safety goggles** and a **filter mask.** Aim the fan across the sediment. Start the fan on its lowest speed. Record your observations in your ScienceLog.

4. Turn the fan to a medium speed and then to a high speed to imitate a desert wind storm. Record your observations at each speed.

5. What is the relationship between the wind speed and the sediment size that is moved?

6. Does the remaining sediment fit the definition of desert pavement?

✓ Self-Check

Why do deflation hollows form in areas where there is little vegetation? *(See page 120 to check your answer.)*

CONNECT TO

LIFE SCIENCE

Each year, equatorial trade winds carry millions of tons of reddish-brown dust from the Sahara Desert to Florida. Sahara dust causes hazy skies in Florida and travels as far as South America, providing nutrients for organisms that live in the rain-forest canopies. Traveling over the Pacific Ocean, yellow dust from Mongolia's Gobi Desert reaches Hawaii, fertilizing iron-deficient regions of the Pacific Ocean. Where the dust settles, plankton populations increase, enriching the food chain. One researcher has linked this phenomenon to global climate change. According to this theory, desertification increases during glacial periods, so more sediment is deposited in the oceans. This deposition encourages plankton growth, which removes CO_2 from the atmosphere, further cooling the planet.

---

2 Teach

Quick Lab

MATERIALS

FOR EACH GROUP:
• dust
• sand
• table
• gravel
• electric fan

Safety Caution: Remind students to review all safety cautions and icons before beginning this activity. Because of the risk of eye injury and particle inhalation, students must wear both protective goggles and a filter mask during this lab.

Answers to QuickLab

5. The faster the wind speed, the larger the sediment that can be moved.

6. Answers may vary. Accept all reasonable responses.

USING THE FIGURE

Direct students' attention to **Figure 9.** Ask them to identify and describe the process that removed the fine sediment and created the desert pavement. (deflation) Encourage them to compare this figure with the desert pavement they created in the QuickLab. Sheltered English

Answer to Self-Check

Deflation hollows form in areas where there is little vegetation because there are no plant roots to anchor the sediment in place.

Teaching Transparency 146 "Saltation"

Directed Reading Worksheet Section 2

2) Teach, *continued*

Answer to APPLY

Answers will vary. Deflation caused the Dust Bowl.

Multicultural CONNECTION

The term *hammada* refers to areas of desert pavement. The term is Arabic in origin, as are many of the words used to describe desert features. This is because many of Earth's deserts are located in Arabic-speaking countries. Challenge students to use the Internet or reference texts to compile a list of terms that describe desert features. (Terms with Arabic origins include *erg,* which describes sandy desert; *reg,* which are loose stones; *barchan,* which are crescent-shaped dunes; and *seif,* which are sword-shaped dunes.)

MEETING INDIVIDUAL NEEDS

Learners Having Difficulty
Students may have difficulty visualizing how rock can be abraded by sand. Demonstrate the abrasiveness of sand by briskly rubbing quartz sandpaper on a softer rock specimen, and allow students to observe the changes. Remind them that sandblasting is used in many industrial applications to "erode" hard surfaces.
Sheltered English

Describing the Dust Bowl

When a long period without rain, known as a *drought,* occurs, areas that are farmed or overgrazed can suffer extensive soil loss and dense dust storms. The removal of plants exposes the soil, making it more vulnerable to wind erosion.

During the 1930s, a section of the Great Plains suffered severe wind erosion and dust storms. This area became known as the *Dust Bowl.* The dust darkened the skies so much that street lights were left on during the day in Midwestern cities.

In areas where the conditions were even worse, people had to string ropes from their houses to their barns so they wouldn't get lost in the dense dust. The dust was so bad that people slept with damp cloths over their face to keep from choking. Describe the major erosional process that caused the Dust Bowl.

Abrasion The grinding and wearing down of rock surfaces by other rock or sand particles is called **abrasion.** Abrasion commonly occurs in areas where there are strong winds, loose sand, and soft rocks. The blowing of millions of sharp sand grains creates a sandblasting effect that helps to erode, smooth, and polish rocks.

Wind-Deposited Materials

Like a stack of papers blowing in the wind, all the material carried by the wind is eventually deposited downwind. The amount and size of particles the wind can carry depend on wind speed. The faster the wind blows, the more material and the heavier the particles it can carry. As wind speed slows, heavier particles are deposited first.

Dunes When the wind hits an obstacle, such as a plant or a rock, it slows down. As the wind slows, it deposits, or drops, the heavier material. As the material collects, it creates an additional obstacle. This obstacle causes even more material to be deposited, forming a mound. Eventually even the original obstacle becomes buried. The mounds of wind-deposited sand are called **dunes.** Dunes are common in deserts and along the shores of lakes and oceans.

64

IS THAT A FACT!

The Yellow River flows across China to the Yellow Sea. The river and sea get their names from the yellow loess that washes off the Gobi Desert and colors the water.

How Dunes Move Dunes tend to move in the direction of strong prevailing winds. Different wind conditions produce dunes in various shapes and sizes. A dune usually has a gently sloped side and a steeply sloped side, or *slip face,* as shown in **Figure 11.** In most cases, the gently sloped side faces the wind. The wind is constantly transporting material up this side of the dune. As sand moves over the crest, or peak, of the dune, it slides down the slip face, creating a steep slope.

Figure 11 *Dunes migrate in the direction of the wind.*

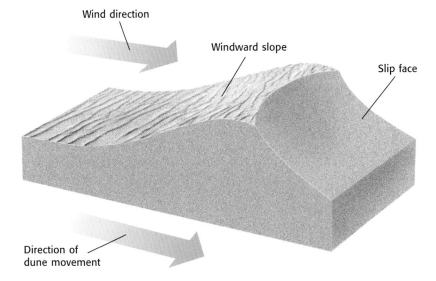

Wind direction

Windward slope

Slip face

Direction of dune movement

Disappearing Dunes and the Desert Tortoise

Dunes provide homes for hundreds of plant and animal species, including the desert tortoise. This tortoise, found in the Mojave and Sonoran Deserts of the southwestern United States, is able to live where ground temperatures are very hot. It escapes the heat by digging burrows in the sand dunes. The desert tortoise has a problem, though. Dune buggies and other motorized vehicles are destroying the dunes. Dunes are easily disturbed and are vulnerable to erosion. Motorized off-road vehicles break down dunes, destroying habitat of the tortoise as well as many other animal and plant species. For this reason, state and federal wildlife and land-management

agencies have taken an active role in helping protect the habitat of the desert tortoise and other sensitive desert species by making some areas off-limits to off-road vehicles.

65

Homework

Create a Diagram Challenge students to make diagrams similar to **Figure 11** illustrating why the gently sloping side of a dune usually faces windward. Have them clearly indicate the direction in which the wind and the sand are moving, and have them write a caption summarizing the phenomenon. (The diagrams should indicate that when moving sand meets an obstacle, it falls to the ground and forms a dune. The wind, however, continues to move forward, causing the sand to form a slope against the obstacle. Wind continues to blow sand up the slope to its apex, whereupon the sand falls, creating a sharper slope on the other side.)

GOING FURTHER

Encourage students to learn about the many different kinds of dunes that can form when windblown sand meets obstacles. Have them make posters illustrating the types of dunes and indicating the wind patterns that create them. Ask students to explore the differences between beach dunes and desert dunes. Display the posters in the classroom.

CONNECT TO
LIFE SCIENCE

Writing Encourage interested students to use library or Internet resources to investigate the living organisms that make sand dunes their home. Have them select one plant or animal and write a brief report describing its habitat, its predators and/or prey, and the adaptations it has made to live in the sand-dune environment. Sheltered English

CROSS-DISCIPLINARY FOCUS

Language Arts Remind students that drought and dust storms plagued the Great Plains in the 1930s. Encourage them to read about the effects of the Dust Bowl in selections from John Steinbeck's *The Grapes of Wrath.* Have them prepare book reports describing how the Joad family was affected economically and psychologically by this event.

Teaching Transparency 147 "Migration of Sand Dunes"

Quiz

1. Why is sand more likely to move by saltation than silt? (Sand is heavier than dust and silt, so as it moves, it tends to bounce along the ground. Silt is light enough to be carried by the wind.)

2. How does the process of deflation form desert pavement? (Deflation lifts and carries away lighter materials, while the heavier stones remain as desert pavement.)

3. Describe how dunes form. (When wind encounters an obstacle, it slows down, depositing some of the heavier material it is carrying. Gradually, this material grows to become a mound and then a dune.)

ALTERNATIVE ASSESSMENT

Concept Mapping Have students use section vocabulary to construct a concept map that explores the ways wind can shape the Earth's surface.

Critical Thinking Worksheet
"A Future in Sand"

internet**connect**

SC*LINKS*
NSTA
TOPIC: Wind Erosion
GO TO: www.scilinks.org
*sci*LINKS NUMBER: HSTE285

Biology
CONNECTION

The sidewinder adder is a poisonous snake that lives in the dunes of the Namib Desert, in southwestern Africa. It is called a sidewinder because of the way it rolls its body to one side as it moves across the sand. This motion allows the snake to move above loose, sliding sand. Its close cousin, the sidewinder rattlesnake, found in the deserts of the southwestern United States, uses a similar motion to move.

internet**connect**

SC*LINKS*
NSTA
TOPIC: Wind Erosion
GO TO: www.scilinks.org
*sci*LINKS NUMBER: HSTE285

Loess Wind can deposit material much finer than sand. Thick deposits of this windblown, fine-grained sediment are known as **loess** (LOH es). Loess feels much like the talcum powder you use after a shower.

Because wind carries fine-grained material much higher and farther than it carries sand, loess deposits are sometimes found far away from their source. Many loess deposits came from glacial sources during the last ice age.

Loess is present in much of the midwestern United States, along the eastern edge of the Mississippi Valley, and in eastern Oregon and Washington. Huge bluffs of loess are found in Mississippi, as shown in **Figure 12.**

Loess deposits can easily be prepared for growing crops and are responsible for the success of many of the grain-growing areas of the world.

Figure 12 *The thick loess deposits found in Mississippi contribute to the state's fertile soil.*

SECTION REVIEW

1. What areas have the greatest amount of wind erosion and deposition? Why?

2. Explain the process of saltation.

3. What is the difference between a dune and a loess deposit?

4. **Analyzing Relationships** Explain the relationship between deflation and dune movement.

▼ *Answers to Section Review*

1. Deserts and coastlines experience the greatest amount of wind erosion and deposition because they are composed of fine, loose material and have little vegetation to anchor the sediment in place.

2. Saltation is the movement of sand-sized particles by bouncing and skipping in the direction that the wind is blowing. The wind lifts sand particles into the air. When the particles land, they hit other particles, causing them to bounce forward.

3. Dunes are made of sand. Loess is made of finer materials that are the size of dust particles. Loess can be carried much higher and farther by wind than can sand.

4. Deflation removes dune sediment. The material is then carried by the wind and redeposited elsewhere, creating another dune.

Terms to Learn

glacier	glacial drift
iceberg	stratified drift
crevasse	till

What You'll Do

- Summarize why glaciers are important agents of erosion and deposition.
- Explain how ice in a glacier flows.
- Describe some of the landforms eroded by glaciers.
- Describe some of the landforms deposited by glaciers.

Erosion and Deposition by Ice

Can you imagine an ice cube the size of a football stadium? Well, glaciers can be even bigger than that. A **glacier** is an enormous mass of moving ice. Because glaciers are very heavy and have the ability to move across the Earth's surface, they are capable of eroding, moving, and depositing large amounts of rock materials. And while you will never see a glacier chilling a punch bowl, you might one day visit some of the spectacular landscapes carved by glacial activity.

Glaciers—Rivers of Ice

Glaciers form in areas so cold that snow stays on the ground year-round. Areas like these, where you can chill a can of juice by simply carrying it outside, are found at high elevations and in polar regions. Because the average temperature is freezing or near freezing, snow piles up year after year. Eventually, the weight of the snow on top causes the deep-packed snow to become ice crystals, forming a giant ice mass. These ice packs then become slow-moving "rivers of ice" as they are set in motion by the pull of gravity on their extraordinary mass.

Alpine Glaciers There are two main types of glaciers, *alpine* and *continental*. **Figure 13** shows an alpine glacier. As you can see, this type of glacier forms in mountainous areas. One common type of alpine glacier is a *valley glacier*. Valley glaciers form in valleys originally created by stream erosion. These glaciers flow slowly downhill, widening and straightening the valleys into broad U-shapes as they travel.

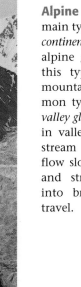

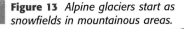

Figure 13 *Alpine glaciers start as snowfields in mountainous areas.*

67

Directed Reading Worksheet Section 3

WEIRD SCIENCE

Glaciers can be very noisy. As they move and stretch, they howl, shriek, pop, groan, and make explosive noises. These sounds can be so loud that they have kept high-altitude mountaineers awake at night!

Focus

Erosion and Deposition by Ice

This section examines how glaciers form and how they shape the Earth's surface. Students will learn to identify different types of glaciers and understand how they move. Finally, students will focus on the changes to the Earth's landscape caused by ice erosion and deposition.

Bellringer

Tell students that 14,000 years ago, much of North America was covered in a thick layer of ice called a *continental glacier,* which moved as far as southern Illinois. Humans were living in North America at the time. Have students imagine they encounter this glacier as an early human, and write a paragraph in their ScienceLog about the experience.

1 Motivate

GROUP ACTIVITY

Glacier Game Give groups of students a world map, two dice, and string to explore the causes and effects of glacial movement. Have each group brainstorm six causes and six effects of glacial movement; these will become 12 possible moves in a glacier game. With each roll of the dice, the group's glacier (represented by a line of string on the map) will either advance or recede. For example, a roll of 5 might mean, "Volcanic eruption triggers global cooling, glacier advances 100 km." The winning group will be the first to have their glacier advance to your hometown.

Answers to Activity

Sample Answer: The iceberg that struck the *Titanic* traveled approximately 2,000 km from Greenland to Newfoundland. Accept all reasonable responses for the mapping exercise.

GUIDED PRACTICE

Have students compare in writing the process of making a snowball with that of forming glacial ice. Students should understand that snowflakes are compressed together when both snowballs and glaciers are formed. When a snowball is made, thermal energy from a person's hands partially melts the snow. As the snowball is squeezed, the snowball becomes denser and harder. Alpine glaciers form in a similar way, but pressure comes from the snow pack above. A cycle of freezing and thawing causes the snow to gradually become glacial ice. Because temperatures always remain below freezing in Antarctica, ice sheets form as lower layers of snow are compressed by the weight of overlying layers. **Sheltered English**

Activity

How far do you think the iceberg that struck the *Titanic* drifted before the two met that fateful night in 1912? Plot on a map of the North Atlantic Ocean the route of the *Titanic* from Southampton, England, to New York. Then plot a possible route of the drifting iceberg from Greenland to where the ship sank, just south of the Canadian island province of Newfoundland.

TRY at HOME

Continental Glaciers Not all glaciers are true "rivers of ice." In fact, some glaciers continue to get larger, spreading across entire continents. These glaciers, called continental glaciers, are huge continuous masses of ice. **Figure 14** shows the largest type of this glacier, a *continental ice sheet*. Ice sheets can cover millions of square kilometers with ice. The continent of Antarctica is almost completely covered by one of the largest ice sheets in the world, as you can see below. This ice sheet is approximately one and a half times the size of the United States. It is so thick—more than 4,000 m in places—that it buries everything but the highest mountain peaks.

Figure 14 *Antarctica contains approximately 91 percent of all the glacial ice on the planet.*

Ice Shelves An area where the ice is attached to the ice sheet but is resting on open water is called an *ice shelf*. The largest ice shelf is the Ross Ice Shelf, shown in **Figure 15,** which is attached to the ice sheet that covers Antarctica. This ice shelf covers an area of ocean about the size of Texas.

Figure 15 *Icebergs break off the Ross Ice Shelf into the Ross Sea.*

Icebergs Large pieces of ice that break off an ice shelf and drift into the ocean are called **icebergs.** The process by which an iceberg forms is called *calving.* Because most of an iceberg is below the surface of the water, it can be a hazard for ships that cannot see how far the iceberg extends. In the North Atlantic Ocean near Newfoundland, the *Titanic* struck an iceberg that calved off the Greenland ice sheet.

68

IS THAT A FACT!

How do snowflakes become massive blocks of glacial ice? As the snow melts and is compacted, the grains become more dense. As snow packs to a greater density, the air spaces among ice crystals are pressed out.

Eventually, the ice recrystallizes to a stage between flakes and ice called *firn*. Over time, with more pressure from overlying layers of snow, the firn will recrystallize again to become glacial ice.

Glaciers on the Move When enough ice builds up on a slope, the ice begins to move downhill. The thickness of the ice and the steepness of the slope determine how fast a glacier will move. Thick glaciers move faster than thin glaciers, and the steeper the slope is, the faster the glacier will move. Glaciers move by two different methods. They move when the weight of the ice causes the ice at the bottom to melt. The water from the melted ice allows the glacier to move forward, like a partially melted ice cube moving across your kitchen counter. Glaciers also move when solid ice crystals within the glacier slip over each other, causing a slow forward motion. This process is similar to placing a deck of cards on a table and then tilting the table. The top cards will slide farther than the lower cards. Like the cards, the upper part, or surface, of the glacier flows faster than the glacier's base.

Crevasses As a glacier flows forward, sometimes crevasses occur. A **crevasse** (kruh VAS), as shown in **Figure 16,** is a large crack that forms where the glacier picks up speed or flows over a high point. Crevasses form because the ice cannot stretch quickly, and it cracks. They can be dangerous for people who are traveling across glaciers because a bridge layer of snow can hide them from view.

Self-Check

How are ice crevasses related to glacier flow?
(See page 120 to check your answer.)

Figure 16 *Crevasses can be dangerous for mountain climbers who must cross glaciers.*

MATH BREAK

Speed of a Glacier

An alpine glacier is estimated to be moving forward at 5 m per day. Calculate how long it will take for the ice to reach a road and campground located 0.5 km from the front of the advancing glacier.

1 km = 1,000 m

IS THAT A FACT!

When metal pipes are drilled through a glacier's layers, they eventually bend in the direction of flow, demonstrating that glacial layers tend to move at different speeds. One cause of this is friction—layers in closest contact with Earth are often slowed by friction.

WEIRD SCIENCE

In September 1991, a melting glacier deposited an unusual load in Italy near its border with Austria: the frozen body of a 5,300-year-old man!

GROUP ACTIVITY

Making Models Divide the class into pairs, and ask each pair to select a landscape feature created by glaciers. Provide modeling clay for students to make a model of the feature. Have each pair present its model to the class and demonstrate how the feature was formed using another color of clay to represent the glacier. **Sheltered English**

Homework

Investigate Your Area If you live in an area that has been affected by glaciers, ask the class to find evidence of glacial erosion and deposition in your area. If not, ask each student to select a landform created by glacial activity and use atlases or other resources to learn where these features are found. For example, the Matterhorn, in Switzerland, is an example of a *horn*. Have students locate their feature on a classroom map and present a short report on what they learned.

CROSS-DISCIPLINARY FOCUS

Language Arts Students will enjoy reading about the adventures of high altitude mountaineers and polar explorers. Have them read selections from the accounts of Antarctic explorers such as Robert Falcon Scott, Ernest Shackleton, or Richard Byrd. Students may also enjoy reading about mountaineers such as Reinhold Messner, Sir Edmund Hillary, or Dr. Johan Reinhard. Students should prepare a presentation for the class about the person they studied, and discuss the explorer's description of glaciers.

Landforms Carved by Glaciers

Alpine glaciers and continental glaciers produce landscapes that are very different from one another. Alpine glaciers carve out rugged features in the mountain rocks through which they flow. Continental glaciers smooth the landscape by scraping and removing features that existed before the ice appeared, flattening even some of the highest mountains. **Figure 17** and **Figure 18** show the very different landscapes that each glacial type produces.

Figure 17
Continental glaciers smooth and flatten the landscape.

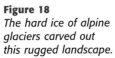

Figure 18
The hard ice of alpine glaciers carved out this rugged landscape.

IS THAT A FACT!

During the last glacial period, the northern part of North America was covered by as much as 4,000 m of ice. The weight of the ice pushed the North American continent down by almost 400 m. Since the ice sheet melted about 10,000 years ago, the continent has risen over 300 m. Parts of the continent are still rising. For example, Maine is rising at a rate of 2mm per year. Parts of Canada are rising even faster. This process is known as isostatic rebound.

Alpine glaciers carve out large amounts of rock material, creating spectacular landforms. These glaciers are responsible for landscapes such as the Rocky Mountains and the Alps. **Figure 19** shows the kind of landscape that is sculpted by alpine glacial erosion and revealed after the ice melts back.

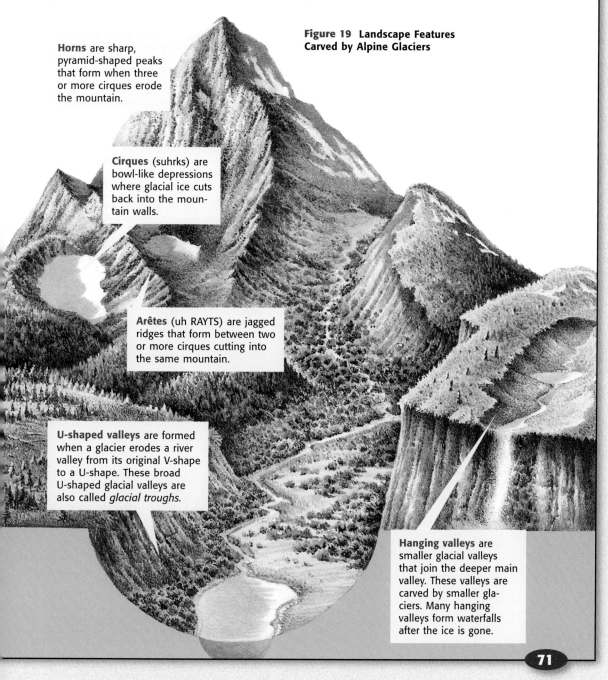

Figure 19 Landscape Features Carved by Alpine Glaciers

Horns are sharp, pyramid-shaped peaks that form when three or more cirques erode the mountain.

Cirques (suhrks) are bowl-like depressions where glacial ice cuts back into the mountain walls.

Arêtes (uh RAYTS) are jagged ridges that form between two or more cirques cutting into the same mountain.

U-shaped valleys are formed when a glacier erodes a river valley from its original V-shape to a U-shape. These broad U-shaped glacial valleys are also called *glacial troughs.*

Hanging valleys are smaller glacial valleys that join the deeper main valley. These valleys are carved by smaller glaciers. Many hanging valleys form waterfalls after the ice is gone.

71

REAL-WORLD CONNECTION

Glaciers throughout the world provide fresh water that helps regulate the flow of large rivers and recharge aquifers. Scientists are concerned, however, that a permanent increase in global temperatures would alter this naturally controlled process. If the 13,800,000 km² Antarctic ice sheet melted, sea level could rise 60 m, with devastating effects. Coastal towns and cities would be flooded, and some islands would disappear. Have students draw a map of what the coastline of the United States would look like if the sea level rose 60 m.

COOPERATIVE LEARNING

Divide the class into small groups, and ask them to imagine an Earth untouched by glaciers. Have them work together to make a poster showing such a planet. Encourage them to consider not only Earth's landscape but also the living things inhabiting it. Ask students to share their posters with the class.

Gliding Glaciers

internet connect

SC⎱LINKS   **TOPIC:** Glaciers
NSTA   **GO TO:** www.scilinks.org
    *sci*LINKS **NUMBER:** HSTE290

Figure 20 *Striations, such as these seen in Central Park, in New York City, are evidence of glacial erosion.*

How did this glacier get into my classroom? To find out more about glaciers and erosion, turn to page 95 of the LabBook.

Striations While many of the erosional features created by glaciers are unique to alpine glaciers, alpine and continental glaciers share some common features. For example, when a glacier erodes the landscape, the glacier picks up rock material and carries it away. This debris is transported on the glacier's surface as well as beneath and within the glacier. Many times, rock material is frozen into the glacier's bottom. As the glacier moves, the rock pieces scrape and polish the surface rock. Larger rocks embedded in the glacier gouge out grooves in the surface rock. As you can see in **Figure 20,** these grooves, called *striations,* help scientists determine the direction of ice flow.

Types of Glacial Deposits

When a glacier melts, all the material it has been carrying is dropped. **Glacial drift** is the general term used to describe all material carried and deposited by glaciers. Glacial drift is divided into two main types, based on whether the material is sorted or unsorted.

Stratified Drift Rock material that has been sorted and deposited in layers by water flowing from the melted ice is called **stratified drift.** Many streams are created by the meltwater from the glacier. These streams carry an abundance of sorted material, which is deposited in front of the glacier in a broad area called an *outwash plain.* Sometimes a block of ice is left in the outwash plain when the glacier retreats. During the time it takes for the ice to melt, sediment builds up around the block of ice. After the ice has melted, a depression called a *kettle* is left. Kettles commonly fill with water, forming a lake or pond, as shown in **Figure 21.**

Figure 21 *Stratified drift is deposited on outwash plains in which kettle lakes are often found.*

CROSS-DISCIPLINARY FOCUS

Geography The spectacular fjords of Norway are underwater valleys carved by glaciers. During the Pleistocene era, glaciers traveled beyond the coastline, digging trenches in the ocean floor. When they retreated to the coast, sea water flooded the valleys that the glaciers had formed. Have groups make a map showing the position of glaciers during different eras of geologic time.

Till Deposits The second type of glacial drift, **till,** is unsorted rock material that is deposited directly by the ice when it melts. *Unsorted* means that the till is made up of different sizes of rock material, ranging from large boulders to fine glacial silt. As a glacier flows, it carries different sizes of rock fragments. When the glacier melts, the unsorted material is deposited on the ground surface.

The most common till deposits are *moraines.* Moraines generally form ridges along the edges of glaciers. They are produced when glaciers carry material to the front of the ice and along the sides of the ice. As the ice melts, the sediment and rock it is carrying are dropped, forming the different types of moraines. The various types of moraines are shown in **Figure 22.**

Medial moraines form when two different valley glaciers with lateral moraines meet.

Lateral moraines form along each side of a glacier.

Ground moraines are the unsorted material left beneath a glacier.

Terminal moraines form when eroded rock material is dropped at the front of the glacier.

Figure 22 *Moraines provide clues to where glaciers once were located.*

SECTION REVIEW

1. How does glaciation change the appearance of mountains?
2. Explain why continental glaciers smooth the landscape and alpine glaciers create a rugged landscape.
3. What do moraines indicate?
4. **Applying Concepts** How can a glacier deposit both sorted and unsorted material?

internet**connect**

*SCI*LINKS.
NSTA

TOPIC: Glaciers
GO TO: www.scilinks.org
*sci*LINKS NUMBER: HSTE290

73

Gravity's Effect on Erosion and Deposition

Focus

Gravity's Effect on Erosion and Deposition

This section introduces gravity as an agent of erosion and deposition. Students learn that mass movements caused by gravity are affected by the material's size, weight, shape, and moisture content and by the slope on which the material rests. Students then learn about landslides, mudslides, and volcanic lahars. The section also examines the effect of slow mass movements, such as creep.

Bellringer

Write the following sentence on the board or overhead projector:

Watch for falling rocks!

Ask students to describe in their ScienceLog places where a warning sign like this would be necessary. Ask students to consider what factors contribute to make a rockfall zone.

1 Motivate

DISCUSSION

Ask students to review the three processes of erosion and deposition that they have learned about so far: shoreline erosion, wind erosion, and glacial erosion. Then ask students how rocks move from mountaintops to valleys. Tell them that gravity is an important force influencing erosion and deposition. Explain that although events like rockfalls and mudslides occur rapidly, most erosion and deposition occurs very slowly as gravity pulls material downward.

Terms to Learn

mass movement   mudflow
rock fall   creep
landslide

What You'll Do

◆ Explain how slope is related to mass movement.
◆ State how gravity affects mass movement.
◆ Describe different types of mass movement.

QuickLab

Angle of Repose

1. Pour a **container** of **dry sand** onto a lab table.
2. With a **protractor,** measure the slope of the sand, or the *angle of repose.*
3. Pour another beaker of sand on top of the first pile.
4. Measure the angle of repose again for the new pile.
5. Which pile is more likely to collapse? Why?

QuickLab

MATERIALS

- beaker
- dry sand
- protractor

Waves, wind, and ice are all agents of erosion and deposition that you can see. And though you can't see it and might not be aware of it, gravity is also an agent of erosion and deposition constantly at work on the Earth's surface. Gravity not only influences the movement of water, such as waves, streams, and ice, but also causes rocks and soil to move downslope. **Mass movement** is the movement of any material, such as rock, soil, or snow, downslope. Mass movement is controlled by the force of gravity and can occur rapidly or slowly.

Angle of Repose

If dry sand is piled up, it will move downhill until the slope becomes stable. The *angle of repose* is the steepest angle, or slope, at which loose material will not slide downslope. This is demonstrated in **Figure 23.** The angle of repose is different for each type of surface material. Characteristics of the surface material, such as its size, weight, shape, and moisture level, determine at what angle the material will move downslope.

Figure 23 *If the slope on which material rests is less than the angle of repose, the material will stay in place. If the slope is greater than the angle of repose, the material will move downslope.*

Answer to QuickLab

5. The second pile is more likely to collapse because it has a steeper slope, which was created by the addition of sediment. The steeper the slope, the more likely mass movement will occur.

Rapid Mass Movement

The most destructive mass movements occur suddenly and rapidly. Rapid mass movement occurs when material, such as rock and soil, moves down-slope quickly. A rapid mass movement can be very dangerous, destroying everything in its path.

Rock Falls While driving along a mountain road, you might have noticed signs that warn of falling rock. A **rock fall** happens when a group of loose rocks falls down a steep slope, as seen in **Figure 24**. Steep slopes are sometimes created to make room for a road in mountainous areas. Loosened and exposed rocks above the road tend to fall as a result of gravity. The rocks in a rock fall can range in size from small fragments to large boulders.

Landslides Another type of rapid mass movement is a *landslide*. A **landslide** is the sudden and rapid movement of a large amount of material downslope. A *slump* is an example of one kind of landslide. Slumping occurs when a block of material moves downslope over a curved surface, as seen in **Figure 25.**

Physics CONNECTION

Gravity is the force of attraction between objects. It is one of the major forces that cause rocks and soil to move from one place to another. The more mass an object has, the more attraction there is between it and other objects.

Figure 24 *If enough rock falls from a mountain, a pile forms at the base of the slope. This pile of rock debris is called a talus slope.*

Figure 25 *A slump is a type of landslide that occurs when a small block of land becomes detached and slides downhill.*

75

2 Teach

Homework

Research Have students research lahars. The Mount St. Helens eruption in 1980, for example, produced tremendously destructive lahars of ash and melted snow. Volcanoes all over the world have created significant lahars or have the potential for major lahars. Have students select one volcano and prepare a hazard report detailing the lahar potential and possible effects on populated areas.

CONNECT TO
PHYSICAL SCIENCE

Use the Teaching Transparency listed below to discuss the force of gravity. Have students write a paragraph in their ScienceLog describing the effects of Earth's gravity on matter. (Students should recognize that gravity is a force of attraction between two masses. The larger the masses are and the closer they are to one another, the stronger gravity is. Because Earth is so massive, it exerts a strong gravitational pull on objects near its surface.)

 Teaching Transparency 220
"The Law of Universal Gravitation"   LINK TO PHYSICAL SCIENCE

 Directed Reading Worksheet Section 4

 internet**connect**
 sciLINKS **TOPIC:** Mass Movement
GO TO: www.scilinks.org
*sci*LINKS **NUMBER:** HSTE295

SCIENTISTS AT ODDS

Long run-out landslides have been the focus of much scientific debate because they appear to defy the laws of physics. Long run-out slides occur when massive amounts of rock, sometimes half a mountainside, suddenly give way. Moving at tremendous speeds (more than 160 km/h), they behave more like liquid than rock.

These landslides can travel 20 times the height that they fall—in 1903, a long run-out landslide in Frank, Canada, traveled 1 km uphill! Some geologists think that these landslides travel on top of a layer of vibrating rocks which act like a conveyer belt, moving materials incredible distances.

ACTIVITY

Making Models Provide students with poster board and markers. Have them each create a diagram of one form of mass movement, such as a landslide, mudslide, or creep. Direct them to label relevant parts of their diagrams and to write captions that explain the phenomenon illustrated.

MEETING INDIVIDUAL NEEDS

Learners Having Difficulty

Ask students to prepare a demonstration that compares mass movements of different materials on varying slopes. Provide a cookie sheet, dry sand, small pebbles, and gravel. Have them raise one end of the cookie sheet about 2 cm. Ask students to use a protractor to measure the angle of repose for each material. Have students repeat the procedure after moistening the materials, and ask them to make conclusions about the effect of water saturation on mass movement.
Sheltered English

REAL-WORLD CONNECTION

Lack of vegetative cover contributes to the frequency and severity of mudslides. Tree roots stabilize the soil and absorb ground water. Deforestation accelerates erosion of slopes. In 1995, there were 260 landslides in British Columbia's Clayquot Sound region during the rainy season. Only 33 were in unlogged areas. As a class, find out more about the connections between large-scale logging operations and recent mudslides. Students may wish to investigate the relationship between deforestation in Central America and the devastating mudslides that followed Hurricane Mitch in 1998.

Mudflows A **mudflow** is a rapid movement of a large mass of mud. Mudflows, which are like giant moving mud pies, occur when a large amount of water mixes with soil and rock. The water causes the slippery mass of mud to flow rapidly downslope. Mudflows most commonly occur in mountainous regions when a long dry season is followed by heavy rains. As you can see in **Figure 26,** a mudflow can carry trees, houses, cars, and other objects that lie in its path.

Figure 26 *This photo shows one of the many mudflows that have occurred in California during rainy winters.*

Lahars The most dangerous mudflows occur as a result of volcanic eruptions. Mudflows of volcanic origin are called *lahars*. Lahars can move at speeds of more than 80 km/h and are as thick as concrete. In mountains with snowy peaks, a volcanic eruption can suddenly melt a great amount of ice, causing a massive and rapid lahar, as shown in **Figure 27.** The water from the ice liquifies the soil and volcanic ash, sending a hot mudflow downslope. Other lahars are caused by heavy rains on volcanic ash.

Figure 27 *This lahar overtook the city of Kyushu, in Japan.*

CONNECT TO
LIFE SCIENCE

In 1980, six successive storms caused devastating mudslides in California. The storms dropped 33 cm of rain, transforming the soil into a sea of oozing mud. Soil on slopes oozed out from under the foundations of houses, sending them crashing into canyons and valleys. Twenty-four people were killed, and millions of dollars in damage was done. Many believe that the mudslides were so massive because the area was recently logged.

Slow Mass Movement

Sometimes you don't even notice mass movement occurring. While rapid mass movements are visible and dramatic, slow mass movements happen a little at a time. However, because slow mass movements occur more frequently, more material is moved collectively over time.

Creep Although most slopes appear to be stable, they are actually undergoing slow mass movement, as shown in **Figure 28.** The extremely slow movement of material downslope is called **creep.** Many factors contribute to creep. Water breaks up rock particles, allowing them to move freely. The roots of growing plants act as a wedge, forcing rocks and soil particles apart. Burrowing animals, such as gophers and groundhogs, loosen rock and soil particles.

Figure 28 *Tilted fence posts and bent tree trunks are evidence that creep is occurring.*

SECTION REVIEW

1. In your own words, explain why slump occurs.
2. What factors increase the potential for mass movement?
3. How do slope and gravity affect mass movement?
4. **Analyzing Relationships** Some types of mass movement are considered dangerous to humans. Which types are most dangerous? Why?

internetconnect

SC*LINKS*
NSTA

TOPIC: Mass Movement
GO TO: www.scilinks.org
*sci*LINKS NUMBER: HSTE295

77

4 Close

Quiz

1. What is the relationship between the angle of repose and the slope required for mass movement to occur? (Mass movement will occur only if the angle of the material is steeper than the angle of repose.)
2. What is creep? (Creep is the slow movement of surface material downslope.)
3. Does slow mass movement or rapid mass movement move more material down a slope? Why? (Slow mass movement, such as creep, occurs more frequently than rapid mass movement, so it moves more material over time.)

ALTERNATIVE ASSESSMENT

Divide the class into small groups, and challenge each group to write a public-service announcement designed to educate the public about the dangers of one form of mass movement. Instruct them to focus on the causes and consequences of these phenomena. Allow time for each group to perform its announcement for the class.

Making Models Lab

Making Models Lab

Dune Movement
Teacher's Notes

Time Required
30 minutes

Lab Ratings
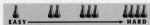
EASY ——→ HARD

TEACHER PREP ▲
STUDENT SET-UP ▲▲
CONCEPT LEVEL ▲
CLEAN UP ▲▲

The materials listed in the student page are enough for two students.

Safety Caution
Remind students to review all safety cautions and icons before beginning this lab activity.

Preparation Notes
You might want to have students do this activity outside, in an area where an electrical outlet is available.

Answers
9. Answers will vary. A typical answer would be about 0.5 to 1.0 cm.
10. Answers will vary. A typical answer would be about 5 to 10 cm.
11. Answers will vary. The sand could be blown until it hits the bag or the dune could move farther. Students may also have predicted that more sand would be blown out of the box.

Dune Movement

Sand bounces along the ground in the direction the wind is blowing in a process called *saltation*. As the sand is blown across a beach, the dunes move. In this activity, you will investigate the effect wind has on a model sand dune.

MATERIALS

- marker
- metric ruler
- shallow cardboard box
- fine sand
- paper bag, large enough to hold half the box
- filter mask
- hair dryer
- watch or clock with a second hand

Procedure

1. Use the marker to draw and label vertical lines 5 cm apart along one side of the box.

2. Fill the box about halfway with sand. Brush the sand into a dune shape about 10 cm from the end of the box.

3. Use the lines you drew along the edge of the box to measure the location of the dune's peak to the nearest centimeter.

4. Slide the box into the paper bag until only about half the box is exposed, as shown below.

5. Put on your safety goggles and filter mask. Hold the hair dryer so that it is level with the peak of the dune and about 10–20 cm from the open end of the box.

6. Turn on the hair dryer at the lowest speed, and direct the air toward the model sand dune for 1 minute.

7. Record the new location of the model dune in your ScienceLog.

8. Repeat steps 5 and 6 three times. After each trial, measure and record the location of the dune's peak.

Analysis

9. How far did the dune move during each trial?

10. How far did the dune move overall?

11. How might the dune's movement be affected if you turned the hair dryer to the highest speed?

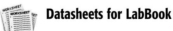
Datasheets for LabBook

CLASSROOM TESTED & APPROVED

Larry Tackett
Andrew Jackson Middle School
Cross Lanes, West Virginia

Discovery Lab

USING SCIENTIFIC METHODS

Creating a Kettle

As glaciers recede, they leave huge amounts of rock material behind. Sometimes receding glaciers form moraines by depositing some of the rock material in ridges. At other times, glaciers leave chunks of ice that form depressions called *kettles*. These depressions may form ponds or lakes. In this activity, you will discover how kettles are formed by creating your own.

MATERIALS

- small tub
- sand
- 4 or 5 ice cubes of various sizes
- metric ruler

Ask a Question

1. How are kettles formed?

Conduct an Experiment

2. Fill the tub three-quarters full with sand.

3. In your ScienceLog, describe the size and shape of the ice cubes.

4. Push the ice cubes to various depths in the sand.

5. Put the tub where it won't be disturbed overnight.

Make Observations

6. Closely observe the sand around the area where you left each ice cube.

7. What happened to the ice cubes?

8. Use a metric ruler to measure the depth and diameter of the indentation left by the ice cubes.

Analyze the Results

9. How does this model relate to the size and shape of a natural kettle?

10. In what ways are your model kettles similar to real ones? How are they different?

Draw Conclusions

11. Based on your model, what can you conclude about the formation of kettles by receding glaciers?

79

CLASSROOM
TESTED & APPROVED
Janel Guse
West Central Middle School
Hartford, South Dakota

WORKSHEET
Datasheets for LabBook

Discovery Lab

Creating a Kettle
Teacher's Notes

Time Required

One 45-minute class period plus 30 minutes during a second day

Lab Ratings

EASY ————————→ HARD

TEACHER PREP
STUDENT SET-UP
CONCEPT LEVEL
CLEAN UP

MATERIALS

The materials listed on the student page are enough for a group of 4–5 students.

Answers

9. Answers will vary. The model is similar. The simulated kettle is the size of the ice cube. A real kettle hole is the size of the block of ice that breaks off a glacier. Its shape is determined by the shape of the ice.

10. Answers will vary. **Similarities:** The ice cube in the lab melted slowly to form a depression. Similarly, blocks of ice left behind by glaciers melt slowly to form kettles. **Differences:** The materials and debris surrounding the model hole are different from those surrounding a real kettle. In addition, a real kettle would not be as uniform as the shape of the model hole.

11. Accept all reasonable responses. Kettles form from the slow melting of ice left behind when a glacier recedes.

Chapter Highlights

Chapter Highlights

VOCABULARY DEFINITIONS

SECTION 1

shoreline the boundary between land and a body of water

beach an area of the shoreline made up of material deposited by waves

longshore current the movement of water near and parallel to the shoreline

SECTION 2

saltation the movement of sand-sized particles by a skipping and bouncing action in the direction the wind is blowing

deflation the lifting and removal of fine sediment by wind

abrasion the grinding and wearing down of rock surfaces by other rock or sand particles

dune a mound of wind-deposited sand

loess thick deposits of windblown, fine-grained sediments

Vocabulary Review Worksheet

Blackline masters of these Chapter Highlights can be found in the **Study Guide.**

SECTION 1

Vocabulary
- **shoreline** *(p. 56)*
- **beach** *(p. 58)*
- **longshore current** *(p. 59)*

Section Notes

- The wind from storms usually produces the large waves that cause shoreline erosion.
- Waves break when they enter shallow water, becoming surf.
- Beaches are made of any material deposited by waves.
- Sandbars and spits are depositional features caused by longshore currents.
- Sea cliffs, sea caves, sea arches, and sea stacks are coastal formations caused by wave erosion.

SECTION 2

Vocabulary
- **saltation** *(p. 62)*
- **deflation** *(p. 63)*
- **abrasion** *(p. 64)*
- **dune** *(p. 64)*
- **loess** *(p. 66)*

Section Notes

- Wind is an important agent of erosion and deposition in deserts and along coastlines.
- Saltation is the process of the wind bouncing sand grains downwind along the ground.
- Deflation is the removal of materials by wind. If deflation removes all fine rock materials, a barren surface called desert pavement is formed.
- Abrasion is the grinding and wearing down of rock surfaces by other rock or sand particles.

- Dunes are formations caused by wind-deposited sand.
- Loess is wind-deposited silt, and it forms soil material good for farming.

☑ Skills Check

Math Concepts

WAVE PERIOD Waves travel in intervals that are usually between 10 and 20 seconds apart. Use the following equation to calculate how many waves reach the shore in 1 minute:

$$\frac{\text{number of waves}}{\text{per minute}} = \frac{60 \text{ seconds}}{\text{waves period (seconds)}}$$

After you find out how many waves reach the shore in 1 minute, you can figure out how many waves occur in an hour or even a day. For example, consider a wave period of 15 seconds. Using the formula above, you find that 4 waves occur in 1 minute. To find out how many waves occur in 1 hour, multiply 4 by 60. To find out how many waves occur in 1 day, multiply 240 by 24.

$$\frac{\text{number of waves}}{\text{per day}} = \frac{60}{15} \times 60 \times 24 = 5{,}760$$

Visual Understanding

U-SHAPED VALLEYS AND MORE Look back at the illustration on page 71 to review the different types of landscape features carved by alpine glaciers.

Lab and Activity Highlights

Dune Movement PG 78

Creating a Kettle PG 79

Gliding Glaciers PG 95

Datasheets for LabBook (blackline masters for these labs)

SECTION 3

Vocabulary
glacier *(p. 67)*
iceberg *(p. 68)*
crevasse *(p. 69)*
glacial drift *(p. 72)*
stratified drift *(p. 72)*
till *(p. 73)*

Section Notes

• Masses of moving ice are called glaciers.

• There are two main types of glaciers—alpine glaciers and continental glaciers.

• Glaciers move when the ice that comes into contact with the ground melts and when ice crystals slip over one another.

• Alpine glaciers produce rugged landscape features, such as cirques, arêtes, and horns.

• Continental glaciers smooth the landscape.

• There are two main types of glacial deposits—stratified drift and till.

• Some of the landforms deposited by glaciers include outwash plains and moraines.

Labs
Gliding Glaciers *(p. 95)*

SECTION 4

Vocabulary
mass movement *(p. 74)*
rock fall *(p. 75)*
landslide *(p. 75)*
mudflow *(p. 76)*
creep *(p. 77)*

Section Notes

• Mass movement is the movement of material downhill due to the force of gravity.

• The angle of repose is the steepest slope at which loose material will remain at rest.

• Rock falls, landslides, and mudflows are all types of rapid mass movement.

• Creep is a type of slow mass movement.

VOCABULARY DEFINITIONS, *continued*

SECTION 3

glacier an enormous mass of moving ice

iceberg a large piece of ice that breaks off an ice shelf and drifts into the ocean

crevasse a large crack that forms where a glacier picks up speed or flows over a high point

glacial drift all material carried and deposited by glaciers

stratified drift rock material that has been sorted and deposited in layers by water flowing from the melted ice of a glacier

till unsorted rock material that is deposited directly by glacial ice when it melts

SECTION 4

mass movement the movement of any material downslope

rock fall a group of loose rocks that fall down a steep slope

landslide a sudden and rapid movement of a large amount of material downslope

mudflow the rapid movement of a large mass of mud/rock and soil mixed with a large amount of water that flows downhill

creep the extremely slow movement of material downslope

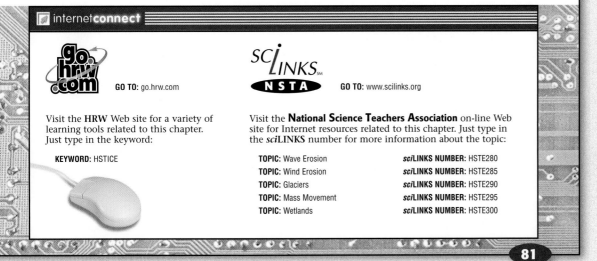

internet connect

go.hrw.com
GO TO: go.hrw.com

Visit the **HRW** Web site for a variety of learning tools related to this chapter. Just type in the keyword:

KEYWORD: HSTICE

SCILINKS
NSTA
GO TO: www.scilinks.org

Visit the **National Science Teachers Association** on-line Web site for Internet resources related to this chapter. Just type in the *sci***LINKS** number for more information about the topic:

TOPIC: Wave Erosion	*sci*LINKS NUMBER: HSTE280
TOPIC: Wind Erosion	*sci*LINKS NUMBER: HSTE285
TOPIC: Glaciers	*sci*LINKS NUMBER: HSTE290
TOPIC: Mass Movement	*sci*LINKS NUMBER: HSTE295
TOPIC: Wetlands	*sci*LINKS NUMBER: HSTE300

81

Lab and Activity Highlights

LabBank

Whiz-Bang Demonstrations
• Between a Rock and a Hard Place
• Rising Mountains

Long-Term Projects & Research Ideas,
Deep in the Mud

Chapter Review

USING VOCABULARY

Explain the difference between the words in the following pairs:

1. shoreline/longshore current
2. beaches/dunes
3. deflation/saltation
4. glacier/loess
5. stratified drift/till
6. mudflow/creep

UNDERSTANDING CONCEPTS

Multiple Choice

7. *Surf* refers to
   a. large storm waves in the open ocean.
   b. giant waves produced by hurricanes.
   c. breaking waves.
   d. small waves on a calm sea.

8. When waves cut completely through a headland, a ___?___ is formed.
   a. sea cave   c. sea stack
   b. sea cliff   d. sea arch

9. A narrow strip of sand that is formed by wave deposition and is connected to the shore is called a ___?___
   a. marine terrace.   c. spit.
   b. sandbar.   d. headland.

10. A wind-eroded depression is called a
    a. dune.   c. deflation hollow.
    b. desert pavement. d. dust bowl.

11. Where is the world's largest ice sheet located?
    a. Greenland
    b. Canada
    c. Alaska
    d. Antarctica

12. The process of calving forms ___?___
    a. continental ice sheets.
    b. icebergs.
    c. U-shaped valleys.
    d. moraines.

13. What term describes all types of glacial deposits?
    a. drift   c. till
    b. loess   d. outwash

14. Which of the following is not a landform created by an alpine glacier?
    a. cirque   c. horn
    b. deflation hollow   d. arête

15. What is the term for a mass movement of volcanic origin?
    a. lahar   c. creep
    b. slump   d. rock fall

16. Which of the following is a slow mass movement?
    a. mudflow   c. creep
    b. landslide   d. rock fall

Short Answer

17. Why do waves break when they get near the shore?

18. What role do storms play in coastal erosion?

19. How do humans increase the erosion caused by dust storms?

Chapter Review Answers

USING VOCABULARY

1. The shoreline is the area where land and a body of water meet. A longshore current is a movement of water near and parallel to the shoreline.
2. A beach is an area of the shoreline made up of material deposited by waves. A dune is a deposit of windblown sand that can be found on a beach.
3. Deflation is the lifting and removal of material by the wind. Saltation is the movement of sand in the direction the wind is blowing by a skipping and bouncing action.
4. A glacier is a large mass of moving ice. Loess is a thick layer of sediment deposited by wind.
5. Stratified drift is sorted glacial drift. Till is unsorted glacial drift.
6. A mudflow is a rapid mass movement. Creep is a slow mass movement.

UNDERSTANDING CONCEPTS

Multiple Choice

7. c
8. c
9. c
10. c
11. d
12. b
13. a
14. b
15. a
16. c

Short Answer

17. When waves reach shallow water, the lower part of the wave is crowded by the ocean floor. The wave becomes taller, eventually growing so tall that it cannot support itself. When it reaches this point, it curls and breaks.

18. Storms generally produce larger waves, which have greater erosive energy. These waves are capable of removing chunks of rock from the coastline.
19. Answers will vary. Sample answer: Sometimes people remove native vegetation from an area to make room for agricultural land. The plants anchor the soil in place. By removing the plants, they make the area more vulnerable to wind erosion.

20. Sand dunes move in the direction of the prevailing winds.
21. Glaciers are effective agents of erosion and deposition because they are very massive and they move across the Earth's surface.
22. Tilted fence posts and bent tree trunks are evidence of creep.

20. In what direction do sand dunes move?

21. Why are glaciers such effective agents of erosion and deposition?

22. List some evidence for creep.

Concept Mapping

23. Use the following terms to create a concept map: deflation, dust storm, saltation, dune, loess.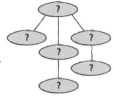

CRITICAL THINKING AND PROBLEM SOLVING

Write one or two sentences to answer the following questions:

24. What role does wind play in the processes of erosion and deposition?

25. What are the main differences between alpine glaciers and continental glaciers?

26. Describe the different types of moraines.

27. What kind of mass movement occurs continuously, day after day? Why can't you see it?

MATH IN SCIENCE

28. While standing on a beach, you can estimate a wave's speed in kilometers per hour. This is done by counting the seconds between each arriving wave crest to determine the wave period and then multiplying the wave period by 3.5. Calculate the speed of a wave with a 10-second period.

INTERPRETING GRAPHICS

The following graph illustrates coastal erosion and deposition occurring at an imaginary beach over a period of 8 years.

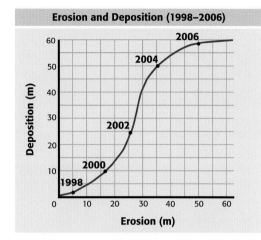

Erosion and Deposition (1998–2006)

29. What is happening to the beach over time?

30. In what year does the amount of erosion that has occurred along the shoreline equal the amount of deposition?

31. Based on the erosion and deposition data for 2000, what might happen to the beach in the years to follow?

Reading Check-up

Take a minute to review your answers to the Pre-Reading Questions found at the bottom of page 54. Have your answers changed? If necessary, revise your answers based on what you have learned since you began this chapter.

Concept Mapping Transparency 12

Blackline masters of this Chapter Review can be found in the **Study Guide.**

Concept Mapping

23. An answer to this exercise can be found in the front of this book.

CRITICAL THINKING AND PROBLEM SOLVING

24. Wind removes sediment through saltation and deflation and deposits it in a different area. Wind also erodes the surface of materials, such as rocks, through abrasion.

25. Alpine glaciers are generally smaller than continental glaciers. Alpine glaciers carve out rugged landscapes, while continental glaciers smooth the landscape. Alpine glaciers form in the mountains, and continental glaciers spread across a continent.

26. Lateral moraines are the till that is deposited on the side of a glacier. Medial moraines form when the lateral moraines of two glaciers are pushed together. Terminal moraines are the till that is deposited at the front of a glacier. Ground moraines are the material left beneath a glacier.

27. Creep happens every day. You can't see it happening because it is a slow mass movement.

MATH IN SCIENCE

28. $10 \times 3.5 = 35$ km/h

INTERPRETING GRAPHICS

29. At first, more soil is being eroded than deposited. The area is losing beach. After 2002, more soil is being deposited than eroded. The beach is getting larger.

30. 2002

31. In 2000, more sand is being eroded than deposited; the coastline is losing land. This would lead you to believe that the coastline would continue shrinking. But by 2002, the coastline stops losing land and begins growing. The most recent data indicates that the coastline will continue growing.

Background

Scientists who first studied the Death Valley boulders suggested that the rocks moved because they were embedded in a rigid ice sheet. The trails of two of the rocks, named Jacki and Julie, showed remarkable congruence when Messina measured the distance between them as they moved. Although their movement across the Racetrack appeared to be a type of synchronized ballet, Messina found that the rocks converged as they twirled around. If the rocks were stuck in ice, they would remain the same distance apart; if the ice sheet shattered, the rocks would diverge. This data allowed Messina to rule out the rigid-ice-sheet theory.

Teaching Strategy

Point out that rain makes the surface of the lake bed slippery, allowing the rocks to be moved more easily. To demonstrate this, you may wish to have students experiment with wet and dry clay to compare the friction between the two surfaces.

Science, Technology, and Society

Boulder Boogie

Karen weighs 320 kg. When no one's looking, she slides around, leaving lots of tracks. But Karen's not a person. In fact, she's not even alive—she's a boulder! Over the years, Karen has moved hundreds of meters across the desert floor. How can a 320 kg rock slide around by itself?

▲ A mystery in Death Valley: What moved this rock across the desert floor?

▲ New technology is helping Paula Messina study the paths of the "dancing rocks."

Slipping and Sliding

Karen is one of the mysterious dancing rocks of Death Valley. These rocks slide around—sometimes together, sometimes alone. There are nearly 200 of them, and they range in size from small to very large. No one has seen them move, but their trails show where they've been.

The rocks are scattered across a dry lake bed, called the Racetrack, in Death Valley, California. The Racetrack is very flat and has almost no plants or wildlife. Several times a year, powerful storms rip across the lake bed, bringing plenty of rain, wind, and sometimes snow. The Racetrack's clay surface becomes slippery, and that's apparently when the rocks dance.

Puzzles and Clues

What could push a 320 kg boulder hundreds of yards across the mud? With the help of technology, scientists like Paula Messina are finally getting some answers. Messina uses a global positioning system (GPS) receiver and a geographic information system (GIS) to study the rocks. Using GPS satellites, Messina is able to map the movements of the rocks. Her measurements are more accurate than ever before. This new device measures the locations within centimeters! A computer equipped with GIS software constructs maps that allow her to study how the rock movement relates to the terrain. Messina's investigations with this equipment have led her to conclude that wind is probably pushing the rocks.

But how does the wind push such massive rocks? Messina thinks the gaps in the mountains at one end of the valley funnel high-speed winds down onto the slippery clay surface, pushing the rocks along. And why do some rocks move while others nearby do not? This mystery will keep Messina returning to Death Valley for years.

Search and Find

▶ Go to the library or the Internet, and research the many uses for GPS devices. Make a list in your ScienceLog of all the uses for GPS devices you find.

Answers to Search and Find

Students' answers will vary but might include one of the following: for marine navigation, for aviation, to find ancient trails, or to find water.

EYE ON THE ENVIRONMENT

Beach Today, Gone Tomorrow

Beaches are fun, right? But what if you went to the coast and found that the road along the beach had washed away? It could happen. In fact, erosion is stripping away beaches from islands and coastlines around the world.

An Island's Beaches

The beaches of Anguilla, a small Caribbean island, are important to the social, economic, and environmental well-being of the island and its inhabitants. Anguilla's sandy shores protect coastal areas from wave action and provide habitats for coastal plants and animals. The shores also provide important recreational areas for tourists and local residents. When Hurricane Luis hit Anguilla in 1995, Barney Bay was completely stripped of sand. But Anguilla's erosion problems started long before Luis hit the island. Normal ocean wave action had already washed away some beaches.

Back in the United States

Louisiana provides a good example of coastal problems in the United States. Louisiana has 40 percent of the nation's coastal wetlands. As important as these wetlands are, parts of the Louisiana coast are disappearing at a rate of 65 to 90 km² per year. That's a football field every 15 minutes! At that rate of erosion, Louisiana's new coastline would be 48 km inland by the year 2040!

Save the Sand

The people of Louisiana and Anguilla have acted to stop the loss of their coastlines. But many of their solutions are only temporary. Waves, storms, and human activity continue to erode coastlines. What can be done about beach erosion?

Scientists know that beaches and wetlands come and go to a certain extent. Erosion is part of a natural cycle. Scientists must first determine how much erosion is normal for

Before

After

▲ *This is what Barney Bay looked like in 1995 before and after Hurricane Luis.*

a particular area and how much is the result of human activities or some unusual process. The next step is to preserve or stabilize existing sand dunes, preserve coastal vegetation, and plant more shrubs, vines, grasses, and trees. The people of Louisiana and Anguilla have learned a lot from their problems and are taking many of these steps to slow further erosion. If steps are taken to protect valuable coastal areas, beaches will be there when you go on vacation.

Extending Your Knowledge

▶ What are barrier islands? How are they related to coastal erosion? On your own, find out more about barrier islands and why it is important to preserve them.

EYE ON THE ENVIRONMENT

Beach Today, Gone Tomorrow

Background

State and federal agencies, as well as concerned Louisiana citizens and businesses, have been taking steps in the right direction to preserve the Louisiana coast. A joint effort is being made to rebuild and restore the wetlands. Rebuilding methods include using sediments from near-shore sandbars to create wetlands and refilling canals, which were previously sending loose sediment directly to the ocean.

*internet*connect

TOPIC: Wetlands
GO TO: www.scilinks.org
*sci*LINKS NUMBER: HSTE300

Answer to Extending Your Knowledge

Barrier islands are long, narrow islands formed by the deposition of sediment. Louisiana's barrier islands formed when the Mississippi River changed course. As one delta was abandoned, a new one formed, and a series of islands developed at the end of the old delta. These islands protect the coast from devastating winds and waves created by offshore storms.

If the barrier islands were to completely disappear, Louisiana could lose an additional 48 km of shoreline. Experts suggest that the most effective means to preserve Louisiana's barrier islands would be to nourish them. Beach nourishment involves the transport of sand from other areas in order to replenish the area lost to erosion.

SAFETY FIRST!

Exploring, inventing, and investigating are essential to the study of science. However, these activities can also be dangerous. To make sure that your experiments and explorations are safe, you must be aware of a variety of safety guidelines.

You have probably heard of the saying, "It is better to be safe than sorry." This is particularly true in a science classroom where experiments and explorations are being performed. Being uninformed and careless can result in serious injuries. Don't take chances with your own safety or with anyone else's.

Following are important guidelines for staying safe in the science classroom. Your teacher may also have safety guidelines and tips that are specific to your classroom and laboratory. Take the time to be safe.

Safety Rules!

Start Out Right

Always get your teacher's permission before attempting any laboratory exploration. Read the procedures carefully, and pay particular attention to safety information and caution statements. If you are unsure about what a safety symbol means, look it up or ask your teacher. You cannot be too careful when it comes to safety. If an accident does occur, inform your teacher immediately, regardless of how minor you think the accident is.

Safety Symbols

All of the experiments and investigations in this book and their related worksheets include important safety symbols to alert you to particular safety concerns. Become familiar with these symbols so that when you see them, you will know what they mean and what to do. It is important that you read this entire safety section to learn about specific dangers in the laboratory.

If you are instructed to note the odor of a substance, wave the fumes toward your nose with your hand. Never put your nose close to the source.

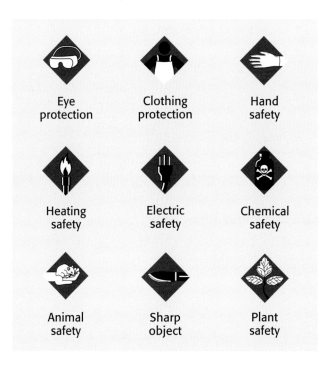

Eye protection

Clothing protection

Hand safety

Heating safety

Electric safety

Chemical safety

Animal safety

Sharp object

Plant safety

Eye Safety

Wear safety goggles when working around chemicals, acids, bases, or any type of flame or heating device. Wear safety goggles any time there is even the slightest chance that harm could come to your eyes. If any substance gets into your eyes, notify your teacher immediately, and flush your eyes with running water for at least 15 minutes. Treat any unknown chemical as if it were a dangerous chemical. Never look directly into the sun. Doing so could cause permanent blindness.

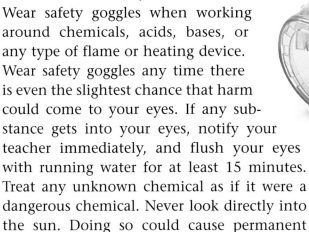

Avoid wearing contact lenses in a laboratory situation. Even if you are wearing safety goggles, chemicals can get between the contact lenses and your eyes. If your doctor requires that you wear contact lenses instead of glasses, wear eye-cup safety goggles in the lab.

Safety Equipment

Know the locations of the nearest fire alarms and any other safety equipment, such as fire blankets and eyewash fountains, as identified by your teacher, and know the procedures for using them.

Be extra careful when using any glassware. When adding a heavy object to a graduated cylinder, tilt the cylinder so the object slides slowly to the bottom.

Neatness

Keep your work area free of all unnecessary books and papers. Tie back long hair, and secure loose sleeves or other loose articles of clothing, such as ties and bows. Remove dangling jewelry. Don't wear open-toed shoes or sandals in the laboratory. Never eat, drink, or apply cosmetics in a laboratory setting. Food, drink, and cosmetics can easily become contaminated with dangerous materials.

Certain hair products (such as aerosol hair spray) are flammable and should not be worn while working near an open flame. Avoid wearing hair spray or hair gel on lab days.

Sharp/Pointed Objects

Use knives and other sharp instruments with extreme care. Never cut objects while holding them in your hands. Place objects on a suitable work surface for cutting.

Heat

Wear safety goggles when using a heating device or a flame. Whenever possible, use an electric hot plate as a heat source instead of an open flame. When heating materials in a test tube, always angle the test tube away from yourself and others. In order to avoid burns, wear heat-resistant gloves whenever instructed to do so.

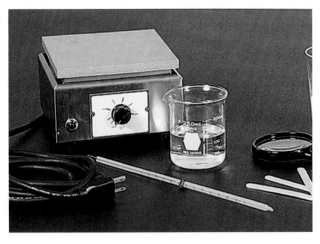

Chemicals

Wear safety goggles when handling any potentially dangerous chemicals, acids, or bases. If a chemical is unknown, handle it as you would a dangerous chemical. Wear an apron and safety gloves when working with acids or bases or whenever you are told to do so. If a spill gets on your skin or clothing, rinse it off immediately with water for at least 5 minutes while calling to your teacher.

Never mix chemicals unless your teacher tells you to do so. Never taste, touch, or smell chemicals unless you are specifically directed to do so. Before working with a flammable liquid or gas, check for the presence of any source of flame, spark, or heat.

Electricity

Be careful with electrical cords. When using a microscope with a lamp, do not place the cord where it could trip someone. Do not let cords hang over a table edge in a way that could cause equipment to fall if the cord is accidentally pulled. Do not use equipment with damaged cords. Be sure your hands are dry and that the electrical equipment is in the "off" position before plugging it in. Turn off and unplug electrical equipment when you are finished.

Animal Safety

Always obtain your teacher's permission before bringing any animal into the school building. Handle animals only as your teacher directs. Always treat animals carefully and with respect. Wash your hands thoroughly after handling any animal.

Plant Safety

Do not eat any part of a plant or plant seed used in the laboratory. Wash hands thoroughly after handling any part of a plant. When in nature, do not pick any wild plants unless your teacher instructs you to do so.

Glassware

Examine all glassware before use. Be sure that glassware is clean and free of chips and cracks. Report damaged glassware to your teacher. Glass containers used for heating should be made of heat-resistant glass.

Orient Yourself!
Teacher's Notes

Time Required

Two 45-minute class periods, one period to learn the use of a compass and the second to follow the orienteering course

Lab Ratings

EASY ———————→ HARD

TEACHER PREP   ▲▲▲
STUDENT SET-UP   ▲
CONCEPT LEVEL   ▲▲▲
CLEAN UP   ▲

MATERIALS

The materials listed on the student page are enough for a group of 3–4 students.

Preparation Notes

Find a suitable outdoor location for a simple orienteering course, and choose five control points for students to map. For example, you may wish to use several pieces of equipment in the playground, the flagpole, a tree, and a small hill. Be sure to mark each control point with either a specific color or a code word that students can collect or note on their maps when they reach each point.

Next, draw a map that includes the control points and the cardinal directions. Label a sixth spot as the starting point. Each group of students will need a copy of this map.

Orient Yourself!

You have been invited to attend an orienteering event with your neighbors. In orienteering events, participants use maps and compasses to find their way along a course. There are several control points that each participant must reach. The object is to reach each control point and then the finish line. Orienteering events are often timed competitions. In order to find the fastest route through the course, the participants must read the map and use their compass correctly. Being the fastest runner does not necessarily guarantee finishing first. You also must choose the most direct route to follow.

Your neighbors participate in several orienteering events each year. They always come home raving about how much fun they had. You would like to join them, but you will need to learn how to use your compass first.

Materials

- magnetic compass
- course map
- ruler
- 2 colored pencils or markers

Procedure

1. Together as a class, go outside to the orienteering course your teacher has made.

2. Hold your compass flat in your hand. Turn the compass until the N is pointing straight in front of you. (The needle in your compass will always point north.) Turn your body until the needle lines up with the N on your compass. You are now facing north.

3. Regardless of which direction you want to face, you should always align the end of the needle with the N on your compass. If you are facing south, the needle will be pointing directly toward your body. When the N is aligned with the needle, the S will be directly in front of you, and you will be facing south.

4. Use your compass to face east. Align the needle with the N. Where is the E? Turn to face that direction. When the needle and the N are aligned and the E is directly in front of you, you are facing east.

5. In an orienteering competition, you will need to know how to determine which direction you are traveling. Now, face any direction you choose.

Before groups begin exploring the orienteering course, have students perform steps 1–6 individually. It may take as much as an entire class period for students to feel confident using a compass.

David Jones
Andrew Jackson Middle School
Cross Lanes, West Virginia

6. Do not move, but rotate the compass to align the needle on your compass with the N. What direction are you facing? You are probably not facing directly north, south, east, or west. If you are facing between north and west, you are facing northwest. If you are facing between north and east, you are facing northeast.

7. Find a partner or partners to follow the course your teacher has made. Get a copy of the course map from your teacher. It will show several control points. You must stop at each one. You will need to follow this map to find your way through the course. Find and stand at the starting point.

8. Face the next control point on your map. Rotate your compass to align the needle on your compass with the N. What direction are you facing?

9. Use the ruler to draw a line on your map between the two control points. Write the direction between the starting point and the next control point on your map.

10. Walk toward the control point. Keep your eyes on the horizon, not on your compass. You might need to go around obstacles such as a fence or building. Use the map to find the easiest way around.

11. Record the color or code word you find at the control point next to the control point symbol on your map.

12. Repeat steps 8–11 for each control point. Follow the points in order as they are labeled. For example, determine the direction from control point 1 to control point 2. Be sure to include the direction between the final control point and the starting point.

Analysis

13. The object of an orienteering competition is to arrive at the finish line first. The maps provided at these events do not instruct the participants to follow a specific path. In one form of orienteering, called "score orienteering," competitors may find the control points in any order. Look at your map. If this course were used for a score-orienteering competition, would you change your route? Explain.

14. If there is time, follow the map again. This time, use your own path to find the control points. Draw this path and the directions on your map in a different color. Do you believe this route was faster? Why?

Going Further

Do some research to find out about orienteering events in your area. The Internet and local newspapers may be good sources for the information. Are there any events that you would like to attend?

Answers

13. Answers will vary. Students should realize that the path shown on the map did not instruct them to follow the most direct route. They should propose a more direct route to follow. Their proposal should include the direction from one control point to the next.

14. This route should be faster. Students should realize that in an orienteering event, participants generally need to determine the quickest route. Some students may also realize that the quickest route is not necessarily the most direct. For example, there may be obstacles in the way (like hills or lakes) that could slow participants down. Orienteering maps include these landmarks.

 Datasheets for LabBook

 Science Skills Worksheet "Designing an Experiment"

91

Topographic Tuber Teacher's Notes

Time Required

One 45-minute class period

Lab Ratings

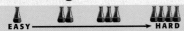

EASY —————————→ HARD

TEACHER PREP 🧪🧪
STUDENT SET-UP 🧪
CONCEPT LEVEL 🧪🧪🧪
CLEAN UP 🧪🧪

MATERIALS

The materials listed on the student page are enough for groups of 2–3 students. Modeling clay may be used in place of the potatoes. Students can mold the clay into a variety of shapes and compare their topographic maps.

Preparation Notes

It may be easier for students to see the waterline if you add a few drops of food coloring to the water before they add it to the container.

Before the activity, select several oddly shaped root vegetables from your local grocery store. Choose vegetables that have varied contour and shape. Sweet potatoes, for example, are available year-round and have many irregular shapes. If you cannot find naturally occurring root vegetables that have odd shapes, shape potatoes with a knife and peeler. You will then need to cut the potatoes in half lengthwise.

Topographic Tuber

Imagine that you live on top of a tall mountain and often look down on the lake below. Every summer, an island appears. You call it Sometimes Island because it goes away again during heavy fall rains. This summer you begin to wonder if you could make a topographic map of Sometimes Island. You don't have fancy equipment to make the map, but you have an idea. What if you place a meterstick with the 0 m mark at the water level in the summer? Then as the expected fall rains come, you could draw the island from above as the water rises. Would this idea really work?

Materials

- clear plastic storage container with transparent lid
- transparency marker
- metric ruler
- potato, cut in half
- water
- tracing paper

Ask a Question

1. How do I make a topographic map?

Conduct an Experiment

2. Place a mark at the storage container's base. Label this mark "0 cm" with a transparency marker.

3. Measure and mark 1 cm increments up the side of the container until you reach the top of the container. Label these marks "1 cm," "2 cm," "3 cm," and so on.

4. The scale for your map will be 1 cm = 10 m. Draw a line 2 cm long in the bottom right-hand corner of the lid. Place hash marks at 0 cm, 1 cm, and 2 cm. Label these marks "0 m," "10 m," and "20 m."

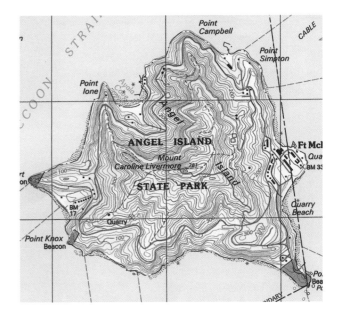

5. Place the potato flat side down in the center of the container.

6. Place the lid on the container, and seal it.

7. Viewing the potato from above, use the transparency marker to trace the outline of the potato where it rests on the bottom of the container. The floor of the container corresponds to the summer water level in the lake.

Lab Notes

Only islands that are at sea level begin at an elevation of 0 m. In order to calculate the elevation of an island that forms in a lake, you must also consider the elevation of the lake.

Michael E. Kral
West Hardin Middle School
Cecilia, Kentucky

8. Label this contour 0 m. (For this activity, assume that the water level in the lake during the summer is the same as sea level.)

9. Pour water into the container until it reaches the line labeled "1 cm."

10. Again place the lid on the container, and seal it. Part of the potato will be sticking out above the water. Viewing the potato from above, trace the part of the potato that touches the top of the water.

11. Label the elevation of the contour line you drew in step 10. According to the scale, the elevation is 10 m.

12. Remove the lid. Carefully pour water into the container until it reaches the line labeled "2 cm."

13. Place the lid on the container, and seal it. Viewing the potato from above, trace the part of the potato that touches the top of the water at this level.

14. Use the scale to calculate the elevation of this line. Label the elevation on your drawing.

15. Repeat steps 12–14, adding 1 cm to the depth of the water each time. Stop when the potato is completely covered.

16. Remove the lid, and set it on a tabletop. Place tracing paper on top of the lid. Trace the contours from the lid onto the paper. Label the elevation of each contour line. Congratulations! You have just made a topographic map!

Analyze the Results

17. What is the contour interval of this topographic map?

18. By looking at the contour lines, how can you tell which parts of the potato are steeper?

19. What is the elevation of the highest point on your map?

Draw Conclusions

20. Do all topographic maps have a 0 m elevation contour line as a starting point? How would this affect a topographic map of Sometimes Island? Explain your answer.

21. Would this method of measuring elevation be an effective way to make a topographic map of an actual area on Earth's surface? Why or why not?

Going Further

Place all of the potatoes on a table or desk at the front of the room. Your teacher will mix up the potatoes as you trade topographic maps with another group. By reading the topographic map you just received, can you pick out the matching potato?

Answers

17. The contour interval of the topographic map is 10 m.

18. Steeper parts of the potato will have contour lines that are closer together.

19. Answers will vary. Elevation is indicated by numbers on the contour lines.

20. No; topographic maps do not necessarily start with a 0 m elevation contour line. It is possible to make a topographic map of Sometimes Island showing its contours but it is impossible to know the island's elevation above sea level.

21. No; flooding an island is not an effective way of mapping it.

Datasheets for LabBook

93

Time Required

Two 45-minute class periods

Lab Ratings

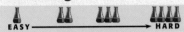

EASY ——————————→ HARD

TEACHER PREP 🧪🧪🧪
STUDENT SET-UP 🧪🧪
CONCEPT LEVEL 🧪
CLEAN UP 🧪🧪

MATERIALS

The materials listed on the student page are enough for 1–3 students. If a freezer is unavailable, students may perform this activity at home.

Safety Caution

Remind students to review all safety cautions and icons before beginning this lab activity. Warn them that when a plastic jar cracks, the pieces could be very sharp.

Preparation Notes

You should perform this lab ahead of time in order to make certain that your plastic jars will break. Be sure to use hard plastic jars. If these jars bend or "give," they may not break. Be sure the students fill the jars to overflowing.

David M. Sparks
Redwater Junior High School
Redwater, Texas

DISCOVERY LAB

Great Ice Escape

Did you know that ice acts as a natural wrecking ball? Even rocks don't stand a chance against the power of ice. When water trapped in rock freezes, a process called *ice wedging* occurs. The water volume increases, and the rock cracks to "get out of the way." This expansion can fragment a rock into several pieces. In this exercise you will see how this natural wrecker works, and you will try to stop the great ice escape.

Ask a Question

1. If a plastic jar is filled with water, is there a way to prevent the jar from breaking when the water freezes?

Conduct an Experiment

2. Fill three identical jars to overflowing with water, and close two of them securely.

3. Measure the height of the water in the unsealed container. Record the height in your ScienceLog.

4. Tightly wrap one of the closed jars with tape, string, or other items to reinforce the jar. These items must be removable. The unwrapped, sealed jar will serve as your control.

5. Place all three jars in resealable sandwich bags, and leave them in the freezer overnight. (Make sure the open jar does not spill.)

6. Remove the jars from the freezer, and carefully remove the wrapping from the reinforced jar.

Make Observations

7. Did your reinforced jar crack? Why or why not?

8. What does each jar look like? Record your observations in your ScienceLog.

9. In your ScienceLog, record the height of the ice in the unsealed jar. How does the new height compare with the height you measured in step 3?

Analyze the Results

10. Do you think it is possible to stop the ice from breaking the sealed jars? Why or why not?

11. How could ice wedging affect soil formation?

Answers

7. Answers will vary.

8. Sample answer: Ice is protruding from the top of the unsealed jar. Both sealed jars are cracked. The unwrapped jar cracked more severely.

9. Answers will vary, but students should observe that the height of the water has increased.

10. No; the expanding ice cannot be confined. If it can shatter a rock, it can break plastic.

11. Ice wedging breaks up large rocks into smaller pieces that are further weathered chemically and physically. The processes of weathering create soil.

Datasheets for LabBook

Gliding Glaciers

A glacier is large moving mass of ice. Glaciers are responsible for shaping many of the Earth's natural features. Glaciers are set in motion by the pull of gravity. As a glacier moves it changes the landscape, eroding the surface over which it passes.

Slip-Sliding Away

The material that is carried by a glacier erodes the Earth's surface, gouging out grooves called *striations.* Different materials have varying effects on the landscape. By creating a model glacier, you will demonstrate the effects of glacial erosion by various materials.

Procedure

1. Fill one margarine container with sand to a depth of 1 cm. Fill another margarine container with gravel to a depth of 1 cm. Leave the third container empty. Fill the containers with water.

2. Put the three containers in a freezer, and leave them overnight.

3. Retrieve the containers from the freezer, and remove the three ice blocks from the containers.

4. Use a rolling pin to flatten the modeling clay.

5. Hold the plain ice block firmly with a towel, and press as you move it along the length of the clay. Do this three times. In your ScienceLog, sketch the pattern the ice block makes in the clay.

6. Repeat steps 4 and 5 with the ice block that contains sand. Sketch the pattern this ice block makes in the clay.

7. Repeat steps 4 and 5 with the ice block that contains gravel. Sketch the pattern this ice block makes in the clay.

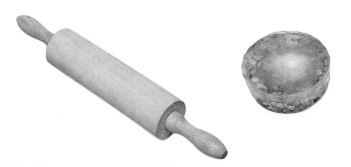

Materials

- 3 empty margarine containers
- sand
- gravel
- metric ruler
- water
- freezer
- rolling pin
- modeling clay
- small towel
- 3 bricks
- 3 pans
- 50 mL graduated cylinder
- timer

95

 **LabBook**

Gliding Glaciers Teacher's Notes

Time Required

Two 45-minute class periods plus a 15-minute activity ahead of time

Lab Ratings

EASY			HARD

TEACHER PREP 🧪🧪
STUDENT SET-UP 🧪
CONCEPT LEVEL 🧪🧪
CLEAN UP 🧪🧪

MATERIALS

The materials listed in the student page are enough for one student or a pair of students. These materials could also be used in larger groups. To reduce the amount of materials, students could use ice cubes and smaller amounts of clay, sand, and gravel.

Preparation Notes

Students should review the entire section on glaciers in this chapter prior to performing this activity. For the second part of the lab, students might have to refreeze ice blocks overnight or make three more ice blocks. If new ice blocks are made, the sand and gravel can be omitted.

Bert Sherwood
Socorro Middle School
El Paso, Texas

Lab Notes

This part of the lab models how the weight of glacial ice causes the ice at the bottom of the glacier to melt. This is one way that glaciers move. You may wish to explain this concept by discussing how ice skates work. Ice skates glide smoothly because they distribute a skater's weight on two thin blades. The weight of a skater applied to such a small surface area causes the ice beneath the blades of the skate to melt and quickly re-freeze. Like glaciers, ice skaters glide on a thin layer of water.

Answers

8. Answers will vary. Small amounts of the surface material may become mixed with the ice.

9. Answers will vary. Small amounts of the material in the ice may be deposited on the clay surface.

10. Answers will vary. Answers may include moraines, striations, and outwash plains.

11. Accept all reasonable, justified answers. Alpine glaciers leave rugged features behind as they flow. Continental glaciers smooth the landscape.

17. The ice block with two bricks on it produced the most water.

18. The bricks represent layers of ice.

19. The bottom of the ice block melted first due to the weight of the bricks on top of it.

20. Students should conclude that glaciers that are heavier melt faster. This, in turn, causes glaciers to move faster.

Analysis

8. Did any material from the clay become mixed with the material in the ice blocks? Explain.

9. Was any material deposited on the clay surface? Explain.

10. What glacial features are represented in your clay model?

11. Compare the patterns formed by the three model glaciers. Do the patterns look like features carved by alpine glaciers or by continental glaciers? Explain.

Slippery When Wet

As the layers of ice build up and the glacier gets larger, the glacier will eventually begin to melt. The water from the melted ice allows the glacier to move forward. In this activity, you'll learn about the effect of pressure on the melting rate of a glacier.

Procedure

12. Place one ice block upside down in each pan.

13. Place one brick on top of one of the ice blocks. Place two bricks on top of another ice block. Leave the third ice block alone.

14. After 15 minutes, remove the bricks from the ice blocks.

Going Further

Replace the clay with different materials, such as soft wood or sand. How does each ice block affect the different surface materials? What types of surfaces do the different materials represent?

15. Measure the amount of water that has melted from each ice block using the graduated cylinder.

16. Record your findings in your ScienceLog.

Analysis

17. Which ice block produced the most water?

18. What did the bricks represent?

19. What part of the ice block melted first? Explain.

20. How could you relate this investigation to the melting rate of glaciers? Explain.

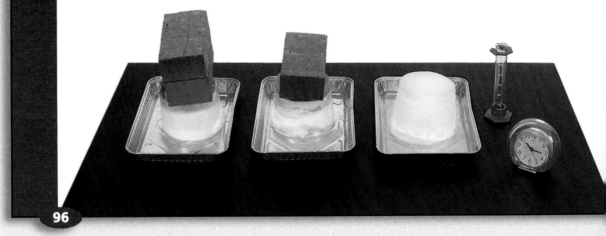

Contents

Concept Mapping: A Way to Bring Ideas Together

What Is a Concept Map?

Have you ever tried to tell someone about a book or a chapter you've just read and found that you can remember only a few isolated words and ideas? Or maybe you've memorized facts for a test and then weeks later discovered you're not even sure what topics those facts covered.

In both cases, you may have understood the ideas or concepts by themselves but not in relation to one another. If you could somehow link the ideas together, you would probably understand them better and remember them longer. This is something a concept map can help you do. A concept map is a way to see how ideas or concepts fit together. It can help you see the "big picture."

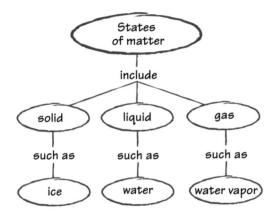

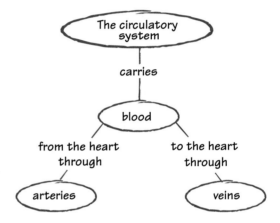

How to Make a Concept Map

❶ Make a list of the main ideas or concepts.

It might help to write each concept on its own slip of paper. This will make it easier to rearrange the concepts as many times as necessary to make sense of how the concepts are connected. After you've made a few concept maps this way, you can go directly from writing your list to actually making the map.

❷ Arrange the concepts in order from the most general to the most specific.

Put the most general concept at the top and circle it. Ask yourself, "How does this concept relate to the remaining concepts?" As you see the relationships, arrange the concepts in order from general to specific.

❸ Connect the related concepts with lines.

❹ On each line, write an action word or short phrase that shows how the concepts are related.

Look at the concept maps on this page, and then see if you can make one for the following terms:

plants, water, photosynthesis, carbon dioxide, sun's energy

One possible answer is provided at right, but don't look at it until you try the concept map yourself.

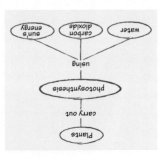

SI Measurement

The International System of Units, or SI, is the standard system of measurement used by many scientists. Using the same standards of measurement makes it easier for scientists to communicate with one another.

SI works by combining prefixes and base units. Each base unit can be used with different prefixes to define smaller and larger quantities. The table below lists common SI prefixes.

SI Prefixes			
Prefix	**Abbreviation**	**Factor**	**Example**
kilo-	k	1,000	kilogram, 1 kg = 1,000 g
hecto-	h	100	hectoliter, 1 hL = 100 L
deka-	da	10	dekameter, 1 dam = 10 m
		1	meter, liter
deci-	d	0.1	decigram, 1 dg = 0.1 g
centi-	c	0.01	centimeter, 1 cm = 0.01 m
milli-	m	0.001	milliliter, 1 mL = 0.001 L
micro-	μ	0.000 001	micrometer, 1 μm = 0.000 001 m

SI Conversion Table		
SI units	**From SI to English**	**From English to SI**
Length		
kilometer (km) = 1,000 m	1 km = 0.621 mi	1 mi = 1.609 km
meter (m) = 100 cm	1 m = 3.281 ft	1 ft = 0.305 m
centimeter (cm) = 0.01 m	1 cm = 0.394 in.	1 in. = 2.540 cm
millimeter (mm) = 0.001 m	1 mm = 0.039 in.	
micrometer (μm) = 0.000 001 m		
nanometer (nm) = 0.000 000 001 m		
Area		
square kilometer (km^2) = 100 hectares	1 km^2 = 0.386 mi^2	1 mi^2 = 2.590 km^2
hectare (ha) = 10,000 m^2	1 ha = 2.471 acres	1 acre = 0.405 ha
square meter (m^2) = 10,000 cm^2	1 m^2 = 10.765 ft^2	1 ft^2 = 0.093 m^2
square centimeter (cm^2) = 100 mm^2	1 cm^2 = 0.155 in.^{2}	1 in.^{2} = 6.452 cm^2
Volume		
liter (L) = 1,000 mL = 1 dm^3	1 L = 1.057 fl qt	1 fl qt = 0.946 L
milliliter (mL) = 0.001 L = 1 cm^3	1 mL = 0.034 fl oz	1 fl oz = 29.575 mL
microliter (μL) = 0.000 001 L		
Mass		
kilogram (kg) = 1,000 g	1 kg = 2.205 lb	1 lb = 0.454 kg
gram (g) = 1,000 mg	1 g = 0.035 oz	1 oz = 28.349 g
milligram (mg) = 0.001 g		
microgram (μg) = 0.000 001 g		

Temperature Scales

Temperature can be expressed using three different scales: Fahrenheit, Celsius, and Kelvin. The SI unit for temperature is the kelvin (K).

Although 0 K is much colder than 0°C, a change of 1 K is equal to a change of 1°C.

Three Temperature Scales

	Fahrenheit	Celsius	Kelvin
Water boils	212°	100°	373
Body temperature	98.6°	37°	310
Room temperature	68°	20°	293
Water freezes	32°	0°	273

Temperature Conversions Table

To convert	Use this equation:	Example
Celsius to Fahrenheit °C ⟶ °F	$°F = \left(\dfrac{9}{5} \times °C\right) + 32$	Convert 45°C to °F. $°F = \left(\dfrac{9}{5} \times 45°C\right) + 32 = 113°F$
Fahrenheit to Celsius °F ⟶ °C	$°C = \dfrac{5}{9} \times (°F - 32)$	Convert 68°F to °C. $°C = \dfrac{5}{9} \times (68°F - 32) = 20°C$
Celsius to Kelvin °C ⟶ K	$K = °C + 273$	Convert 45°C to K. $K = 45°C + 273 = 318 \text{ K}$
Kelvin to Celsius K ⟶ °C	$°C = K - 273$	Convert 32 K to °C. $°C = 32 \text{ K} - 273 = -241°C$

Measuring Skills

Using a Graduated Cylinder

When using a graduated cylinder to measure volume, keep the following procedures in mind:

1 Make sure the cylinder is on a flat, level surface.

2 Move your head so that your eye is level with the surface of the liquid.

3 Read the mark closest to the liquid level. On glass graduated cylinders, read the mark closest to the center of the curve in the liquid's surface.

Using a Meterstick or Metric Ruler

When using a meterstick or metric ruler to measure length, keep the following procedures in mind:

1 Place the ruler firmly against the object you are measuring.

2 Align one edge of the object exactly with the zero end of the ruler.

3 Look at the other edge of the object to see which of the marks on the ruler is closest to that edge. **Note:** Each small slash between the centimeters represents a millimeter, which is one-tenth of a centimeter.

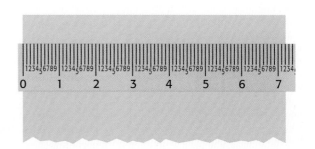

Using a Triple-Beam Balance

When using a triple-beam balance to measure mass, keep the following procedures in mind:

1 Make sure the balance is on a level surface.

2 Place all of the countermasses at zero. Adjust the balancing knob until the pointer rests at zero.

3 Place the object you wish to measure on the pan. **Caution:** Do not place hot objects or chemicals directly on the balance pan.

4 Move the largest countermass along the beam to the right until it is at the last notch that does not tip the balance. Follow the same procedure with the next-largest countermass. Then move the smallest countermass until the pointer rests at zero.

5 Add the readings from the three beams together to determine the mass of the object.

6 When determining the mass of crystals or powders, use a piece of filter paper. First find the mass of the paper. Then add the crystals or powder to the paper and re-measure. The actual mass of the crystals or powder is the total mass minus the mass of the paper. When finding the mass of liquids, first find the mass of the empty container. Then find the mass of the liquid and container together. The mass of the liquid is the total mass minus the mass of the container.

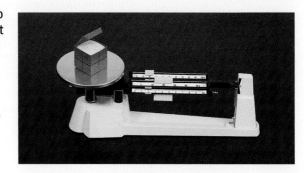

Scientific Method

The series of steps that scientists use to answer questions and solve problems is often called the **scientific method.** The scientific method is not a rigid procedure. Scientists may use all of the steps or just some of the steps of the scientific method. They may even repeat some of the steps. The goal of the scientific method is to come up with reliable answers and solutions.

Six Steps of the Scientific Method

1 **Ask a Question** Good questions come from careful **observations.** You make observations by using your senses to gather information. Sometimes you may use instruments, such as microscopes and telescopes, to extend the range of your senses. As you observe the natural world, you will discover that you have many more questions than answers. These questions drive the scientific method.

Questions beginning with *what, why, how,* and *when* are very important in focusing an investigation, and they often lead to a hypothesis. (You will learn what a hypothesis is in the next step.) Here is an example of a question that could lead to further investigation.

Question: How does acid rain affect plant growth?

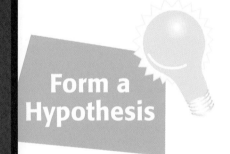

2 **Form a Hypothesis** After you come up with a question, you need to turn the question into a **hypothesis.** A hypothesis is a clear statement of what you expect the answer to your question to be. Your hypothesis will represent your best "educated guess" based on your observations and what you already know. A good hypothesis is testable. If observations and information cannot be gathered or if an experiment cannot be designed to test your hypothesis, it is untestable, and the investigation can go no further.

Here is a hypothesis that could be formed from the question, "How does acid rain affect plant growth?"

Hypothesis: Acid rain causes plants to grow more slowly.

Notice that the hypothesis provides some specifics that lead to methods of testing. The hypothesis can also lead to predictions. A **prediction** is what you think will be the outcome of your experiment or data collection. Predictions are usually stated in an "if . . . then" format. For example, **if** meat is kept at room temperature, **then** it will spoil faster than meat kept in the refrigerator. More than one prediction can be made for a single hypothesis. Here is a sample prediction for the hypothesis that acid rain causes plants to grow more slowly.

Prediction: If a plant is watered with only acid rain (which has a pH of 4), then the plant will grow at half its normal rate.

3 **Test the Hypothesis** After you have formed a hypothesis and made a prediction, you should test your hypothesis. There are different ways to do this. Perhaps the most familiar way is to conduct a **controlled experiment.** A controlled experiment tests only one factor at a time. A controlled experiment has a **control group** and one or more **experimental groups.** All the factors for the control and experimental groups are the same except for one factor, which is called the **variable.** By changing only one factor, you can see the results of just that one change.

Sometimes, the nature of an investigation makes a controlled experiment impossible. For example, dinosaurs have been extinct for millions of years, and the Earth's core is surrounded by thousands of meters of rock. It would be difficult, if not impossible, to conduct controlled experiments on such things. Under such circumstances, a hypothesis may be tested by making detailed observations. Taking measurements is one way of making observations.

Test the Hypothesis

4 **Analyze the Results** After you have completed your experiments, made your observations, and collected your data, you must analyze all the information you have gathered. Tables and graphs are often used in this step to organize the data.

Analyze the Results

5 **Draw Conclusions** Based on the analysis of your data, you should conclude whether or not your results support your hypothesis. If your hypothesis is supported, you (or others) might want to repeat the observations or experiments to verify your results. If your hypothesis is not supported by the data, you may have to check your procedure for errors. You may even have to reject your hypothesis and make a new one. If you cannot draw a conclusion from your results, you may have to try the investigation again or carry out further observations or experiments.

Draw Conclusions

Do they support your hypothesis?

No

Yes

6 **Communicate Results** After any scientific investigation, you should report your results. By doing a written or oral report, you let others know what you have learned. They may want to repeat your investigation to see if they get the same results. Your report may even lead to another question, which in turn may lead to another investigation.

Communicate Results

Scientific Method in Action

The scientific method is not a "straight line" of steps. It contains loops in which several steps may be repeated over and over again, while others may not be necessary. For example, sometimes scientists will find that testing one hypothesis raises new questions and new hypotheses to be tested. And sometimes, testing the hypothesis leads directly to a conclusion. Furthermore, the steps in the scientific method are not always used in the same order. Follow the steps in the diagram below, and see how many different directions the scientific method can take you.

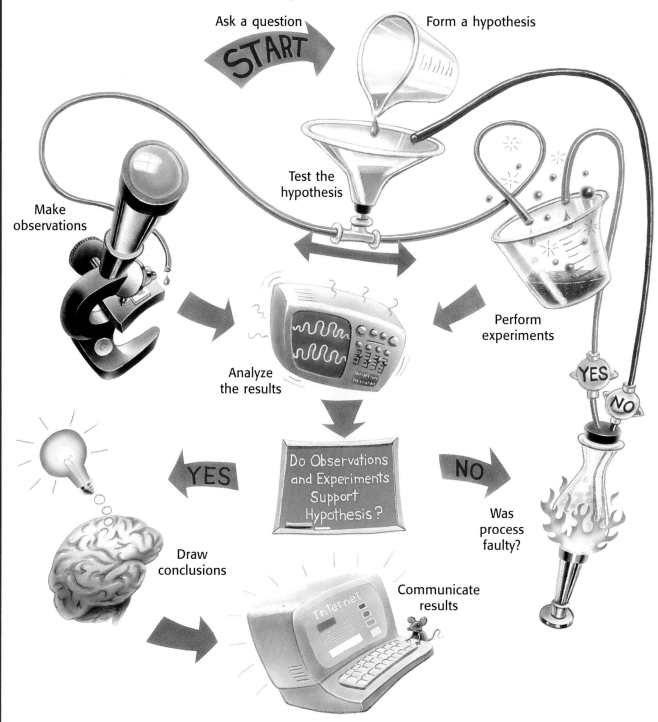

Ask a question

START

Form a hypothesis

Make observations

Test the hypothesis

Perform experiments

YES

NO

Analyze the results

Do Observations and Experiments Support Hypothesis?

YES

NO

Draw conclusions

Was process faulty?

Communicate results

Making Charts and Graphs

Circle Graphs

A circle graph, or pie chart, shows how each group of data relates to all of the data. Each part of the circle represents a category of the data. The entire circle represents all of the data. For example, a biologist studying a hardwood forest in Wisconsin found that there were five different types of trees. The data table at right summarizes the biologist's findings.

Wisconsin Hardwood Trees	
Type of tree	**Number found**
Oak	600
Maple	750
Beech	300
Birch	1,200
Hickory	150
Total	3,000

How to Make a Circle Graph

1 In order to make a circle graph of this data, first find the percentage of each type of tree. To do this, divide the number of individual trees by the total number of trees and multiply by 100.

$$\frac{600 \text{ oak}}{3{,}000 \text{ trees}} \times 100 = 20\%$$

$$\frac{750 \text{ maple}}{3{,}000 \text{ trees}} \times 100 = 25\%$$

$$\frac{300 \text{ beech}}{3{,}000 \text{ trees}} \times 100 = 10\%$$

$$\frac{1{,}200 \text{ birch}}{3{,}000 \text{ trees}} \times 100 = 40\%$$

$$\frac{150 \text{ hickory}}{3{,}000 \text{ trees}} \times 100 = 5\%$$

2 Now determine the size of the pie shapes that make up the chart. Do this by multiplying each percentage by 360°. Remember that a circle contains 360°.

$20\% \times 360° = 72°$ $25\% \times 360° = 90°$
$10\% \times 360° = 36°$ $40\% \times 360° = 144°$
$5\% \times 360° = 18°$

3 Then check that the sum of the percentages is 100 and the sum of the degrees is 360.

$20\% + 25\% + 10\% + 40\% + 5\% = 100\%$
$72° + 90° + 36° + 144° + 18° = 360°$

4 Use a compass to draw a circle and mark its center.

5 Then use a protractor to draw angles of 72°, 90°, 36°, 144°, and 18° in the circle.

6 Finally, label each part of the graph, and choose an appropriate title.

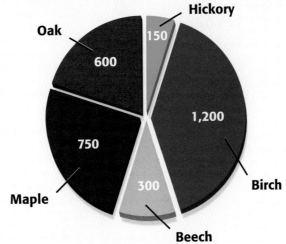

A Community of Wisconsin Hardwood Trees

Line Graphs

Population of Appleton, 1900–2000	
Year	**Population**
1900	1,800
1920	2,500
1940	3,200
1960	3,900
1980	4,600
2000	5,300

Line graphs are most often used to demonstrate continuous change. For example, Mr. Smith's science class analyzed the population records for their hometown, Appleton, between 1900 and 2000. Examine the data at left.

Because the year and the population change, they are the *variables*. The population is determined by, or dependent on, the year. Therefore, the population is called the **dependent variable**, and the year is called the **independent variable**. Each set of data is called a **data pair**. To prepare a line graph, data pairs must first be organized in a table like the one at left.

How to Make a Line Graph

❶ Place the independent variable along the horizontal (x) axis. Place the dependent variable along the vertical (y) axis.

❷ Label the x-axis "Year" and the y-axis "Population." Look at your largest and smallest values for the population. Determine a scale for the y-axis that will provide enough space to show these values. You must use the same scale for the entire length of the axis. Find an appropriate scale for the x-axis too.

❸ Choose reasonable starting points for each axis.

❹ Plot the data pairs as accurately as possible.

❺ Choose a title that accurately represents the data.

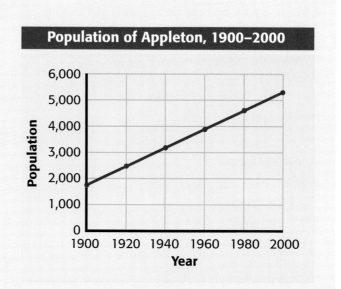

Population of Appleton, 1900–2000

How to Determine Slope

Slope is the ratio of the change in the y-axis to the change in the x-axis, or "rise over run."

❶ Choose two points on the line graph. For example, the population of Appleton in 2000 was 5,300 people. Therefore, you can define point *a* as (2000, 5,300). In 1900, the population was 1,800 people. Define point *b* as (1900, 1,800).

❷ Find the change in the y-axis.
(y at point *a*) − (y at point *b*)
5,300 people − 1,800 people = 3,500 people

❸ Find the change in the x-axis.
(x at point *a*) − (x at point *b*)
2000 − 1900 = 100 years

❹ Calculate the slope of the graph by dividing the change in y by the change in x.

$$\text{slope} = \frac{\text{change in } y}{\text{change in } x}$$

$$\text{slope} = \frac{3,500 \text{ people}}{100 \text{ years}}$$

slope = 35 people per year

In this example, the population in Appleton increased by a fixed amount each year. The graph of this data is a straight line. Therefore, the relationship is **linear.** When the graph of a set of data is not a straight line, the relationship is **nonlinear.**

Using Algebra to Determine Slope

The equation in step 4 may also be arranged to be:

$$y = kx$$

where y represents the change in the y-axis, k represents the slope, and x represents the change in the x-axis.

$$\text{slope} = \frac{\text{change in } y}{\text{change in } x}$$

$$k = \frac{y}{x}$$

$$k \times x = \frac{y \times x}{x}$$

$$kx = y$$

Bar Graphs

Bar graphs are used to demonstrate change that is not continuous. These graphs can be used to indicate trends when the data are taken over a long period of time. A meteorologist gathered the precipitation records at right for Hartford, Connecticut, for April 1–15, 1996, and used a bar graph to represent the data.

Precipitation in Hartford, Connecticut April 1–15, 1996

Date	Precipitation (cm)	Date	Precipitation (cm)
April 1	0.5	April 9	0.25
April 2	1.25	April 10	0.0
April 3	0.0	April 11	1.0
April 4	0.0	April 12	0.0
April 5	0.0	April 13	0.25
April 6	0.0	April 14	0.0
April 7	0.0	April 15	6.50
April 8	1.75		

How to Make a Bar Graph

❶ Use an appropriate scale and a reasonable starting point for each axis.

❷ Label the axes, and plot the data.

❸ Choose a title that accurately represents the data.

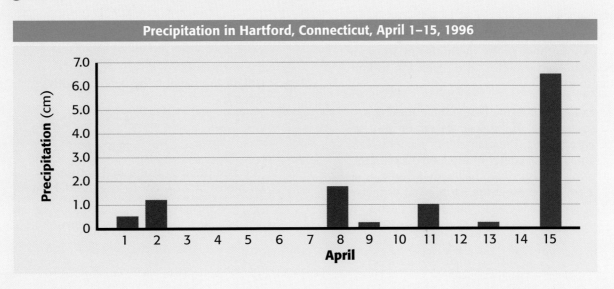

Precipitation in Hartford, Connecticut, April 1–15, 1996

Math Refresher

Science requires an understanding of many math concepts. The following pages will help you review some important math skills.

Averages

An **average,** or **mean,** simplifies a list of numbers into a single number that *approximates* their value.

Example: Find the average of the following set of numbers: 5, 4, 7, and 8.

Step 1: Find the sum.

$$5 + 4 + 7 + 8 = 24$$

Step 2: Divide the sum by the amount of numbers in your set. Because there are four numbers in this example, divide the sum by 4.

$$\frac{24}{4} = 6$$

The average, or mean, is **6.**

Ratios

A **ratio** is a comparison between numbers, and it is usually written as a fraction.

Example: Find the ratio of thermometers to students if you have 36 thermometers and 48 students in your class.

Step 1: Make the ratio.

$$\frac{36 \text{ thermometers}}{48 \text{ students}}$$

Step 2: Reduce the fraction to its simplest form.

$$\frac{36}{48} = \frac{36 \div 12}{48 \div 12} = \frac{3}{4}$$

The ratio of thermometers to students is **3 to 4,** or $\frac{3}{4}$. The ratio may also be written in the form 3:4.

Proportions

A **proportion** is an equation that states that two ratios are equal.

$$\frac{3}{1} = \frac{12}{4}$$

To solve a proportion, first multiply across the equal sign. This is called cross-multiplication. If you know three of the quantities in a proportion, you can use cross-multiplication to find the fourth.

Example: Imagine that you are making a scale model of the solar system for your science project. The diameter of Jupiter is 11.2 times the diameter of the Earth. If you are using a plastic-foam ball with a diameter of 2 cm to represent the Earth, what diameter does the ball representing Jupiter need to be?

$$\frac{11.2}{1} = \frac{x}{2 \text{ cm}}$$

Step 1: Cross-multiply.

$$\frac{11.2}{1} \times \frac{x}{2}$$

$$11.2 \times 2 = x \times 1$$

Step 2: Multiply.

$$22.4 = x \times 1$$

Step 3: Isolate the variable by dividing both sides by 1.

$$x = \frac{22.4}{1}$$

$$x = 22.4 \text{ cm}$$

You will need to use a ball with a diameter of **22.4 cm** to represent Jupiter.

Percentages

A **percentage** is a ratio of a given number to 100.

> **Example:** What is 85 percent of 40?

Step 1: Rewrite the percentage by moving the decimal point two places to the left.

$$.85$$

Step 2: Multiply the decimal by the number you are calculating the percentage of.

$$0.85 \times 40 = 34$$

85 percent of 40 is **34.**

Decimals

To **add** or **subtract decimals,** line up the digits vertically so that the decimal points line up. Then add or subtract the columns from right to left, carrying or borrowing numbers as necessary.

> **Example:** Add the following numbers: 3.1415 and 2.96.

Step 1: Line up the digits vertically so that the decimal points line up.

$$\begin{array}{r} 3.1415 \\ +\ 2.96 \\ \hline \end{array}$$

Step 2: Add the columns from right to left, carrying when necessary.

$$\begin{array}{r} {}^{1\ 1} \\ 3.1415 \\ +\ 2.96 \\ \hline 6.1015 \end{array}$$

The sum is **6.1015.**

Fractions

Numbers tell you how many; **fractions** tell you *how much of a whole.*

> **Example:** Your class has 24 plants. Your teacher instructs you to put 5 in a shady spot. What fraction does this represent?

Step 1: Write a fraction with the total number of parts in the whole as the denominator.

$$\frac{?}{24}$$

Step 2: Write the number of parts of the whole being represented as the numerator.

$$\frac{5}{24}$$

$\frac{5}{24}$ of the plants will be in the shade.

Reducing Fractions

It is usually best to express a fraction in simplest form. This is called *reducing* a fraction.

> **Example:** Reduce the fraction $\frac{30}{45}$ to its simplest form.

Step 1: Find the largest whole number that will divide evenly into both the numerator and denominator. This number is called the greatest common factor (GCF).

factors of the numerator 30: 1, 2, 3, 5, 6, 10, **15,** 30

factors of the denominator 45: 1, 3, 5, 9, **15,** 45

Step 2: Divide both the numerator and the denominator by the GCF, which in this case is 15.

$$\frac{30}{45} = \frac{30 \div 15}{45 \div 15} = \frac{2}{3}$$

$\frac{30}{45}$ reduced to its simplest form is $\frac{2}{3}$.

Adding and Subtracting Fractions

To **add** or **subtract fractions** that have the **same denominator,** simply add or subtract the numerators.

Examples:

$$\frac{3}{5} + \frac{1}{5} = ? \quad \text{and} \quad \frac{3}{4} - \frac{1}{4} = ?$$

Step 1: Add or subtract the numerators.

$$\frac{3}{5} + \frac{1}{5} = \frac{4}{} \quad \text{and} \quad \frac{3}{4} - \frac{1}{4} = \frac{2}{}$$

Step 2: Write the sum or difference over the denominator.

$$\frac{3}{5} + \frac{1}{5} = \frac{4}{5} \quad \text{and} \quad \frac{3}{4} - \frac{1}{4} = \frac{2}{4}$$

Step 3: If necessary, reduce the fraction to its simplest form.

$\frac{4}{5}$ cannot be reduced, and $\frac{2}{4} = \frac{1}{2}$.

To **add** or **subtract fractions** that have **different denominators,** first find the least common denominator (LCD).

Examples:

$$\frac{1}{2} + \frac{1}{6} = ? \quad \text{and} \quad \frac{3}{4} - \frac{2}{3} = ?$$

Step 1: Write the equivalent fractions with a common denominator.

$$\frac{3}{6} + \frac{1}{6} = ? \quad \text{and} \quad \frac{9}{12} - \frac{8}{12} = ?$$

Step 2: Add or subtract.

$$\frac{3}{6} + \frac{1}{6} = \frac{4}{6} \quad \text{and} \quad \frac{9}{12} - \frac{8}{12} = \frac{1}{12}$$

Step 3: If necessary, reduce the fraction to its simplest form.

$\frac{4}{6} = \frac{2}{3}$, and $\frac{1}{12}$ cannot be reduced.

Multiplying Fractions

To **multiply fractions,** multiply the numerators and the denominators together, and then reduce the fraction to its simplest form.

Example:

$$\frac{5}{9} \times \frac{7}{10} = ?$$

Step 1: Multiply the numerators and denominators.

$$\frac{5}{9} \times \frac{7}{10} = \frac{5 \times 7}{9 \times 10} = \frac{35}{90}$$

Step 2: Reduce.

$$\frac{35}{90} = \frac{35 \div 5}{90 \div 5} = \frac{7}{18}$$

Dividing Fractions

To **divide fractions,** first rewrite the divisor (the number you divide *by*) upside down. This is called the reciprocal of the divisor. Then you can multiply and reduce if necessary.

Example:

$$\frac{5}{8} \div \frac{3}{2} = ?$$

Step 1: Rewrite the divisor as its reciprocal.

$$\frac{3}{2} \rightarrow \frac{2}{3}$$

Step 2: Multiply.

$$\frac{5}{8} \times \frac{2}{3} = \frac{5 \times 2}{8 \times 3} = \frac{10}{24}$$

Step 3: Reduce.

$$\frac{10}{24} = \frac{10 \div 2}{24 \div 2} = \frac{5}{12}$$

Scientific Notation

Scientific notation is a short way of representing very large and very small numbers without writing all of the place-holding zeros.

Example: Write 653,000,000 in scientific notation.

Step 1: Write the number without the place-holding zeros.

653

Step 2: Place the decimal point after the first digit.

6.53

Step 3: Find the exponent by counting the number of places that you moved the decimal point.

6.53000000

The decimal point was moved eight places to the left. Therefore, the exponent of 10 is positive 8. Remember, if the decimal point had moved to the right, the exponent would be negative.

Step 4: Write the number in scientific notation.

$$6.53 \times 10^8$$

Area

Area is the number of square units needed to cover the surface of an object.

Formulas:

Area of a square = side × side

Area of a rectangle = length × width

Area of a triangle = $\frac{1}{2}$ × base × height

Examples: Find the areas.

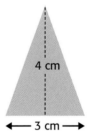

Triangle

Area = $\frac{1}{2}$ × base × height

Area = $\frac{1}{2}$ × 3 cm × 4 cm

Area = **6 cm²**

4 cm

3 cm

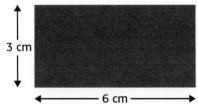

3 cm

6 cm

Rectangle

Area = length × width

Area = 6 cm × 3 cm

Area = **18 cm²**

3 cm

3 cm

Square

Area = side × side

Area = 3 cm × 3 cm

Area = **9 cm²**

Volume

Volume is the amount of space something occupies.

Formulas:

Volume of a cube =
side × side × side

Volume of a prism =
area of base × height

Examples:
Find the volume
of the solids.

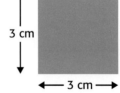

4 cm   3 cm

5 cm

Cube

Volume = side × side × side

Volume = 4 cm × 4 cm × 4 cm

Volume = **64 cm³**

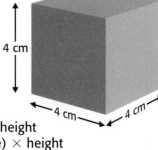

4 cm

4 cm   4 cm

Prism

Volume = area of base × height

Volume = (area of triangle) × height

Volume = $\left(\frac{1}{2} \times 3 \text{ cm} \times 4 \text{ cm} \right) \times 5$ cm

Volume = 6 cm² × 5 cm

Volume = **30 cm³**

Properties of Common Minerals

Silicate Minerals

Mineral	Color	Luster	Streak	Hardness
Beryl	deep green, pink, white, bluish green, or light yellow	vitreous	none	7.5–8
Chlorite	green	vitreous to pearly	pale green	2–2.5
Garnet	green or red	vitreous	none	6.5–7.5
Hornblende	dark green, brown, or black	vitreous or silky	none	5–6
Muscovite	colorless, gray, or brown	vitreous or pearly	white	2–2.5
Olivine	olive green	vitreous	none	6.5–7
Orthoclase	colorless, white, pink, or other colors	vitreous to pearly	white or none	6
Plagioclase	blue gray to white	vitreous	white	6
Quartz	colorless or white; any color when not pure	vitreous or waxy	white or none	7

Nonsilicate Minerals

Native Elements

Mineral	Color	Luster	Streak	Hardness
Copper	copper-red	metallic	copper-red	2.5–3
Diamond	pale yellow or colorless	vitreous	none	10
Graphite	black to gray	submetallic	black	1–2

Carbonates

Mineral	Color	Luster	Streak	Hardness
Aragonite	colorless, white, or pale yellow	vitreous	white	3.5–4
Calcite	colorless or white to tan	vitreous	white	3

Halides

Mineral	Color	Luster	Streak	Hardness
Fluorite	light green, yellow, purple, bluish green, or other colors	vitreous	none	4
Halite	colorless or gray	vitreous	white	2.5–3

Oxides

Mineral	Color	Luster	Streak	Hardness
Hematite	reddish brown to black	metallic to earthy	red to red-brown	5.6–6.5
Magnetite	iron black	metallic	black	5–6

Sulfates

Mineral	Color	Luster	Streak	Hardness
Anhydrite	colorless, bluish, or violet	vitreous to pearly	white	3–3.5
Gypsum	white, pink, gray, or colorless	vitreous, pearly, or silky	white	1–2.5

Sulfides

Mineral	Color	Luster	Streak	Hardness
Galena	lead gray	metallic	lead gray to black	2.5
Pyrite	brassy yellow	metallic	greenish, brownish, or black	6–6.5

Density (g/cm³)	Cleavage, Fracture, Special Properties	Common Uses
2.6–2.8	1 cleavage direction; irregular fracture; some varieties fluoresce in ultraviolet light	gemstones, ore of the metal beryllium
2.6–3.3	1 cleavage direction; irregular fracture	
4.2	no cleavage; conchoidal to splintery fracture	gemstones, abrasives
3.2	2 cleavage directions; hackly to splintery fracture	
2.7–3	1 cleavage direction; irregular fracture	electrical insulation, wallpaper, fireproofing material, lubricant
3.2–3.3	no cleavage; conchoidal fracture	gemstones, casting
2.6	2 cleavage directions; irregular fracture	porcelain
2.6–2.7	2 cleavage directions; irregular fracture	ceramics
2.6	no cleavage; conchoidal fracture	gemstones, concrete, glass, porcelain, sandpaper, lenses
8.9	no cleavage; hackly fracture	wiring, brass, bronze, coins
3.5	4 cleavage directions; irregular to conchoidal fracture	gemstones, drilling
2.3	1 cleavage direction; irregular fracture	pencils, paints, lubricants, batteries
2.95	2 cleavage directions; irregular fracture; reacts with hydrochloric acid	minor source of barium
2.7	3 cleavage directions; irregular fracture; reacts with weak acid, double refraction	cements, soil conditioner, whitewash, construction materials
3.2	4 cleavage directions; irregular fracture; some varieties fluoresce or double refract	hydrochloric acid, steel, glass, fiberglass, pottery, enamel
2.2	3 cleavage directions; splintery to conchoidal fracture; salty taste	tanning hides, fertilizer, salting icy roads, food preservation
5.25	no cleavage; splintery fracture; magnetic when heated	iron ore for steel, gemstones, pigments
5.2	2 cleavage directions; splintery fracture; magnetic	iron ore
2.89–2.98	3 cleavage directions; conchoidal to splintery fracture	soil conditioner, sulfuric acid
2.2–2.4	3 cleavage directions; conchoidal to splintery fracture	plaster of Paris, wallboard, soil conditioner
7.4–7.6	3 cleavage directions; irregular fracture	batteries, paints
5	no cleavage; conchoidal to splintery fracture	dyes, inks, gemstones

Glossary

A

abrasion the grinding and wearing down of rock surfaces by other rock or sand particles (31, 64)

acid precipitation precipitation that contains acids due to air pollution (33)

aerial photograph a photograph taken from the air (13)

arête (uh RAYT) a jagged ridge that forms between two or more cirques cutting into the same mountain (71)

azimuthal (AZ i MYOOTH uhl) **projection** a map projection that is made by transferring the contents of the globe onto a plane (12)

B

beach an area of the shoreline made up of material deposited by waves (58)

bedrock the layer of rock beneath soil (39)

C

cardinal directions north, south, east, and west (5)

chemical weathering the chemical breakdown of rocks and minerals into new substances (33)

cirque (suhrk) a bowl-like depression where glacial ice cuts back into mountain walls (71)

conic (KAHN ik) **projection** a map projection that is made by transferring the contents of the globe onto a cone (12)

contour interval the difference in elevation between one contour line and the next (17)

contour lines lines that connect points of equal elevation (16)

creep the extremely slow movement of material downslope (77)

crevasse (kruh VAS) a large crack that forms where a glacier picks up speed or flows over a high point (69)

D

deflation the lifting and removal of fine sediment by wind (63)

differential weathering the process by which softer, less weather-resistant rocks wear away, leaving harder, more weather-resistant rocks behind (36)

dune a mound of wind-deposited sand (64)

E

elevation the height of an object above sea level; the height of surface landforms above sea level (16)

equator a circle halfway between the poles that divides the Earth into the Northern and Southern Hemispheres (7)

erosion the removal and transport of material by wind, water, or ice (44)

G

glacial drift all material carried and deposited by glaciers (72)

glacier an enormous mass of moving ice (67)

H

hanging valley a small glacial valley that joins the deeper main valley (71)

horn a sharp, pyramid-shaped peak that forms when three or more cirques erode a mountain (71)

humus (HYU muhs) very small particles of decayed plant and animal material in soil (39)

hypothesis a possible explanation or answer to a question (102)

I

iceberg a large piece of ice that breaks off an ice shelf and drifts into the ocean (68)

ice wedging the mechanical weathering process in which water seeps into cracks in rock, freezes, then expands, opening the cracks even wider (30)

index contour a darker, heavier contour line that is usually every fifth line and is labeled by elevation (17)

L

landslide a sudden and rapid movement of a large amount of material downslope (75)

latitude the distance north or south from the equator; measured in degrees (7, 000)

leaching the process by which rainwater dissolves and carries away the minerals and nutrients in topsoil (40)

loess (LOH ES) thick deposits of windblown, fine-grained sediments (66)

longitude the distance east or west from the prime meridian; measured in degrees (8)

longshore current the movement of water near and parallel to the shoreline (59)

M

magnetic declination the angle of correction for the difference between geographic north and magnetic north (6)

map a model or representation of the Earth's surface (4)

mass movement the movement of any material downslope (74)

mechanical weathering the breakdown of rock into smaller pieces by physical means (30)

Mercator projection a map projection that results when the contents of the globe are transferred onto a cylinder (11)

mudflow the rapid movement of a large mass of mud/rock and soil mixed with a large amount of water that flows downhill (76)

O

oxidation a chemical reaction in which an element combines with oxygen to form an oxide (35)

P

parent rock rock that is the source of soil (39)

prime meridian the line of longitude that passes through Greenwich, England; represents 0° longitude (8)

R

reference point a fixed place on the Earth's surface from which direction and location can be described (5)

relief the difference in elevation between the highest and lowest points of an area being mapped (17)

remote sensing gathering information about something without actually being nearby (13)

residual soil soil that remains above the bedrock from which it formed (39)

rock fall a group of loose rocks that fall down a steep slope (75)

S

saltation the movement of sand-sized particles by a skipping and bouncing action in the direction the wind is blowing (62)

scientific method a series of steps that scientists use to answer questions and solve problems (102)

shoreline the boundary between land and a body of water (56)

soil a loose mixture of small mineral fragments and organic material (39)

soil conservation the various methods by which humans take care of the soil (43)

stratified drift rock material that has been sorted and deposited in layers by water flowing from the melted ice of a glacier (72)

T

till unsorted rock material that is deposited directly by glacial ice when it melts (73)

topographic map a map that shows the surface features of the Earth (16)

topsoil the top layer of soil that generally contains humus (40)

transported soil soil that has been blown or washed away from its parent rock (39)

true north the geographic North Pole (6)

U

U-shaped valley a valley that forms when a glacier erodes a river valley from its original V shape to a U shape (71)

W

weathering the breakdown of rock into smaller and smaller pieces by mechanical or chemical means (30)

Index

A **boldface** number refers to an illustration on that page.

Credits

ILLUSTRATIONS

All work, unless otherwise noted, contributed by Holt, Rinehart & Winston.

Table of Contents: T/K

Scope and Sequence: T11, Paul DiMare, T13, Dan Stuckenschneider/Uhl Studios, Inc.

Chapter One: Page 5 John White/The Neis Group; 7, 8(cl), 9, 11, 12, MapQuest.com.

Chapter Two: Page 30(b), Uhl Studios, Inc.; 32(br), Will Nelson/Sweet Reps; 35(r), 36(b), 37, Stephen Durke/Washington Artists; 40(r), Will Nelson/Sweet Reps; 44(c), 48(c), Uhl Studios, Inc.

Chapter Three: Page 58(bl), Uhl Studios, Inc.; 60, 61, Mike Wepple/PAS Group; 62(b), Dean Flemming; 62(cl), Keith Locke; 65(c), Uhl Studios, Inc.; 71(b), Robert Haynes; 83, Sidney Jablonski.

Appendix: Page 100(c), Terry Guyer; 104(b), Mark Mille/Sharon Langley.

PHOTOGRAPHY

Cover and Title Page: Carr Clifton/Minden Pictures

Sam Dudgeon/HRW Photo: Page vii-1, 5(tr), 8(bl), 16(tr), 52, 74, 86, 88(tr,cl,br), 89(tl,b), 90, 91, 93, 94, 95, 96, 97(c,b), 101(br).

Table of Contents: v(tr), Tim Laman/Adventure Photo & Film; v(cr), Jeff Foott/Tom Stack & Associates; v(b), Sam Dudgeon/HRW Photo; vi(tl), Mickey Gibson/Animals Animals; vi(cl), Don Herbert/FPG International; vi(bl), C. Campbell/Corbis; vii(tr), Royal Geographical Society, London, UK/The Bridgeman Art Library; vii(cr), Index Stock; vii(br), Aaron Chang/The Stock Market.

Scope and Sequence: T8(l), Lee F. Snyder/Photo Researchers, Inc.; T8(r), Stephen Dalton/Photo Researchers, Inc.; T10, E. R. Degginger/Color-Pic, Inc., T12(l), Rob Matheson/The Stock Market

Master Materials List: T25(bl, c, br), Image ©2001 PhotoDisc; T26(t), Sam Dudgeon/HRW Photo; T26(br, cl), Image ©2001 PhotoDisc

Chapter One: Page 4, Royal Geographical Society, London, UK/The Bridgeman Art Library; 5, John White/The Neis Group; 6, Tom Van Sant/The Stock Market; 10, Andy Christiansen/HRW Photo; 13, Ariel Images, Incorporated and SOVIN-FORMSPUTNIK; 14-15, The American Map Corporation/ADC The Map People; 16(bl), 17, 18, USGS; 23, Tom Van Sant/The Stock Market; 24(tr), Vladimir Pcholkin/FPG International; 24(bl), Andy Christiansen/HRW Photo; 25, USGS; 26, JPL/NASA; 27(tl), Andy Christiansen/HRW Photo; 27(br), Courtesy Lower Colorado River Authority.

Chapter Two: Page 30, SuperStock; 31(tr), Index Stock Photography; 31(bl), Martin G. Miller/Visuals Unlimited; 31(br), Grant Heilman/Grant Heilman Photography; 32, Andy Christiansen/HRW Photo; 33(bl), Alan Briere/ SuperStock; 34(t), Laurence Parent; 34(bl) C. Campbell/Corbis; 35(cl), SuperStock; 35(cr), Bob Krueger/Photo Researchers, Inc.; 36(b), B. Ross/Corbis; 38, Steven Ferry/HRW Photo; 39(cl), John D. Cunningham/Visuals Unlimited; 39(br), The G.R. "Dick" Roberts Photo Library; 41(t), Tim Laman/Adventure Photo & Film; 41(b), Bill Ross/Corbis; 42(t), Bruce Coleman, Inc.; 42(b), Lee Rentz/Bruce Coleman, Inc.; 43(cl), Grant Heilman Photography, Inc.; 43(bl), Charlton Photos Inc.; 43(br), Tom Walker/Stone; 44(bl), Grant Heilman/Grant Heilman Photography, Inc.; 45(l), Paul Chelsey/Stone; 45(r), Mark Lewis and Adventure Photo & Film; 48(b), Bill Ross/Corbis; 49, Tom Walker/Stone; 50(tr), SuperStock; 50(bl), Mark A. Schneider/Visuals Unlimited; 53, Donald Specker/ Animals Animals/Earth Scenes.

Chapter Three: Page 56, Aaron Chany/The Stock Market; 57(t), Philip Long/Stone; 57(b), SuperStock; 58(tl), Don Herbert/FPG International; 58(cl), Jonathan Weston/Adventure Photo & Film; 58(bl), SuperStock; 59(tl,br), Index Stock; 59(bl), NASA; 60(tl), The G.R. "Dick" Roberts Photo Library; 60(bl), Jeff Foott/Tom Stack & Associates; 60(br), Jeff Foott/DRK Photo; 62(tl), Breck P. Kent; 61(tr), John S. Shelton; 63(t), Breck P. Kent; 63(b), Tom Bean; 64, Brown Brothers; 65(b), Mickey Gibson/Animals Animals/Earth Scenes; 66(l), Michael Fogden/Bruce Coleman Inc.; 66(r), Walter H. Hodge/Peter Arnold, Inc.; 67, Tom Bean/DRK Photo; 68(t), Barbara Gerlach/DRK Photo; 68(b), Tui De Roy/The Roving Tortoise Photography; 69, Didier Givois/Photo Researchers, Inc.; 70(t), Glenn M. Oliver; 70(b), Stone; 72(t), Breck P. Kent; 72(b), 73, Tom Bean; 75(t), A.J. Copley/Visuals Unlimited; 75(b), The G.R. "Dick" Roberts Photo Library; 76(t), Jebb Harris/Orange County Register/SAPA; 76(b), Mike Yamashita/Woodfin Camp & Associates; 77(l), John D. Cunningham/ Visuals Unlimited; 77(r), J & B Photographers/Animals Animals/Earth Scenes; 80, Brown Brothers; 81, Didier Givois, Agence Vandy Stadt/Photo Researchers, Inc.; 82(t), Face of the Earth/Tom Stack & Associates; 82(b), Michael Fredericks/ Animals Animals/Earth Scenes; 84(l), Richard Sisk/Panoramic Images; 84(r), Jane Stevens/Courtesy Paula Messina; 85(tr,cr), Gillian Chambers/ UNESCO/Coping with beach erosion, Coastal Management Sourcebooks.

LabBook/Appendix: "LabBook Header", "L", Corbis Images; "a", Letraset Phototone; "b", and "B", HRW; "o", and "k", images ©2001 PhotoDisc/HRW; Page 87(tr), John Langford/HRW Photo; 87(cl), Michelle Bridwell/HRW Photo; 87(br), Image ©2001 PhotoDisc, Inc./HRW; 88(bl), Stephanie Morris/HRW Photo; 89(tr), Jana Birchum/HRW Photo; 92, USGS; 97(t), Laurence Parent; 101(tr), Peter Van Steen/HRW Photo.

Feature Borders: Unless otherwise noted below, all images copyright ©2001 PhotoDisc/HRW. "Across the Sciences" 52, all images by HRW; "Careers" 27, sand bkgd and Saturn, Corbis Images; DNA, Morgan Cain & Associates; scuba gear, ©1997 Radlund & Associates for Artville; "Eye on the Environment" 53, 85, clouds and sea in bkgd, HRW; bkgd grass, red eyed frog, Corbis Images; hawks, pelican, Animals Animals/Earth Scenes; rat, Visuals Unlimited/ John Grelach; endangered flower, Dan Suzio/Photo Researchers, Inc.; "Science Technology and Society" 26, 84, robot, Greg Geisler.

Self-Check Answers

Chapter 1—Maps as Models of the Earth

Page 6: The Earth rotates around the geographic poles.

Page 17: If the lines are close together, then the mapped area is steep. If the lines are far apart, the mapped area has a gradual slope or is flat.

Chapter 2—Weathering and Soil Formation

Page 32: Water expands as it freezes. This expansion exerts a force great enough to crack rock.

Chapter 3—Agents of Erosion and Deposition

Page 57: A large wave has more erosive energy than a small wave because a large wave releases more energy when it breaks.

Page 63: Deflation hollows form in areas where there is little vegetation because there are no plant roots to anchor the sediment in place.

Page 69: When a moving glacier picks up speed or flows over a high point, a crevasse may form. This occurs because the ice cannot stretch quickly while it is moving, and it cracks.